Cyanobacteria - Recent Advances and New Perspectives

Edited by Archana Tiwari

Published in London, United Kingdom

Cyanobacteria - Recent Advances and New Perspectives
http://dx.doi.org/10.5772/intechopen.102152
Edited by Archana Tiwari

Contributors
Shailendra Yadav, Shilpa Chandra, Mamta Awasthi, Amardeep Yadav, Avinash Kumar, Suchitra Rakesh, Arosha Vaniyankandy, Bobita Ray, Subburamu Karthikeyan, Arumugam Nagarajan, Moahansundaram Yogeswar, Natarajan Valle, Stephen Opoku-Duah, Dennis Johnson, Dan Blair, Jeff Dimick, Eman Elagamey, Magdi A. E. Abdellatef, Hassan E. Flefel, Luisa María Munera Porras, Alejandra Sandoval Valencia, Lisseth Dahiana Salas, María Alejandra Pérez Gutiérrez, Leonardo Alberto Ríos-Osorio, Bahareh Nowruzi, Archana Tiwari, Rashi Tyagi, Pankaj Kumar Singh

First published in London, United Kingdom, 2023 by IntechOpen
IntechOpen is the global imprint of INTECHOPEN LIMITED, registered in England and Wales, registration number: 11086078, 5 Princes Gate Court, London, SW7 2QJ, United Kingdom

British Library Cataloguing-in-Publication Data
A catalogue record for this book is available from the British Library

Additional hard and PDF copies can be obtained from orders@intechopen.com

Cyanobacteria - Recent Advances and New Perspectives
Edited by Archana Tiwari
p. cm.
Print ISBN 978-1-80356-461-6
Online ISBN 978-1-80356-462-3
eBook (PDF) ISBN 978-1-80356-463-0

Meet the editor

Archana Tiwari is a Professor at Amity Institute of Biotechnology, Amity University, Noida, India. She holds a Ph.D. from the University of Allahabad and is a gold medalist and distinction holder in botany. Her key research interests include phycoprospecting diatoms for wastewater remediation and high-value products. Her research has been published in many international journals, and she has written ten books and several book chapters. She received the Researcher of the Year Award in 2015 from Noida International University; the Distinguished Scientist Award in 2016 from the Society for Recent Development in Agriculture; a Woman Scientist Award in 2021 from the Biotech Research Society, India in recognition of her outstanding contributions in algal biotechnology; and an ESDA fellowship in 2022 from the Environment and Social Development Association.

Contents

Preface

This book highlights significant advances in the field of cyanobacteria.

Chapter 1 focuses on picocyanobacteria in surface water bodies, while Chapter 2 discusses the removal of microcystins from drinking water by electrocoagulation. Chapter 3 looks at cyanobacteria as a source of antioxidants, and Chapter 4 explores the role of cyanobacteria as an effective tool in sustainable agriculture. In Chapter 5 cyanobacteria as a source of biodegradable plastics is considered, and Chapter 6 explains the thermochemical conversion of algal-based biofuel. Chapter 7 describes the green synthesis of silver nanoparticles from cyanobacteria and Chapter 8 discusses the potential of cyanobacteria natural products for antibiotic drug discovery.

I hope the book will be useful to researchers, academics and students across the world. I am grateful to all the authors for their valuable contributions and to IntechOpen, especially Author Service Manager Ms. Martina Scerbe for her patience and support. Heartfelt thanks to my angels Aarushi and Anviti for their unconditional support and to my mother Ms Vidya Pandey for encouraging me to work hard and always chase dreams.

This book is dedicated to my father in heaven.

Archana Tiwari
Professor,
Amity Institute of Biotechnology,
Amity University
Noida, India

Chapter 1

Picocyanobacteria in Surface Water Bodies

Alejandra Sandoval Valencia, Lisseth Dahiana Salas, María Alejandra Pérez Gutiérrez, Luisa María Munera Porras and Leonardo Alberto Ríos-Osorio

Abstract

Cyanobacterial harmful algal blooms (CyanoHABs) in lentic, low tidal water bodies with high concentrations of easily assimilated nutrients have generated worldwide concern. However, CyanoHABs often formed from a variety of lesser-known taxa, such as nanocyanobacteria and picocyanobacteria, which are characterized as numerous and ubiquitous in diverse environments. Studies indicate that some taxa of picocyanobacteria can produce toxins. However, their identification through conventional methods is limited by their size and physiological plasticity, recently molecular methods have been chosen for more reliable results. This systematic review aims to summarize the results of original research articles on predominant picocyanobacteria in surface water bodies collected in indexed journal articles and gray literature. The methodology used consisted of searching for original publications in 3 specific databases and one general, using thesauri and free terms; the articles were filtered by previously defined inclusion and exclusion criteria. Thirty-four articles were selected and analyzed. The results show that the predominant picocyanobacteria in freshwater systems belong to the genus *Synechococcus*, reported in oligotrophic systems and capable of producing cyanotoxins. Likewise, from 2015 to 2019, the largest number of publications on this topic was obtained, mainly in countries such as China and the United States, which invest in research resources.

Keywords: cyanobacteria, cyanoHAB, freshwater, picocyanobacteria, surface water

1. Introduction

In recent years, the excessive growth of phytoplanktonic organisms in reservoirs, lagoons, and in general, in lentic, low-tide water bodies with a high concentration of phosphate and nitrogenous nutrients, which are easily assimilated, has generated worldwide concern [1, 2]. Physical factors such as temperature, solar radiation, wind, rain, water column stratification, water flow, or biological interactions with other organisms, among others, play an important role in the so-called cyanobacterial harmful blooms (CyanoHAB) [3, 4].

CyanoHABs can be defined as events in which visually noticeable turbidity of the water occurs due to a rapid accumulation of cyanobacterial cells, often at the water surface, but sometimes deeper in the water column [3, 5]. These blooms have the potential to generate a variety of adverse effects due to their ability to produce toxins [6, 7] that, in turn, cause negative impacts on animals, including humans, aquatic ecosystems, the economy, drinking water supply, real property values, and recreational activities, including swimming and commercial and recreational fishing [8, 9].

CyanoHAB-forming cyanobacteria are often accompanied by a variety of lesser-known taxa that contribute greatly to the total cyanoHAB biomass, such as nanocyanobacteria and picocyanobacteria [10]; e.g., Vardaka et al. [11] have described blooms composed of multiple species.

Picocyanobacteria are bacteria that play a key role in primary production and dominate phytoplankton biomass in both oligotrophic and eutrophic waters [12, 13]. They are the smallest cell-sized, most numerous, and ubiquitous cyanobacteria in freshwater, marine, and even in environments with high salt concentrations [14]. Among the cyanobacteria are the so-called planktonic picocyanobacteria; microorganisms that are part of the smallest aquatic plankton and are often associated with various species [15]. In freshwater, the main representatives are the genera *Synechococcus*, *Cyanobium,* and *Synechocystis*, and in brackish water, *Synechococcus* and *Prochlorococcus* predominate [13, 16].

Although this group of microorganisms is ubiquitous and causes environmental concerns, it is still understudied [13]. Much research continues to use microscopy techniques that require long processing times and can produce erroneous results [17] since picocyanobacteria are difficult to observe and most of the time are found forming groups or present diverse biological forms ranging from single cells to microcolonies [18]; besides, their physiological or epigenetic plasticity means that cyanobacteria with the same genotype can appear very different due to the external factors to which they are influenced [19]. This is determined by the growth conditions, adaptations, and expansion of the cells in response to the stay in complex communities and fluctuating environments [20].

Recently, attempts have been made to study picocyanobacteria through molecular techniques by amplification and sequencing of the 16S rRNA gene or next-generation sequencing (NGS), which allows obtaining results quickly, with high sensitivity and high detection efficiency [21]. However, research aimed at describing picocyanobacteria present in surface waters is atomized, moreover, it is limited and there are no current review articles focused on this. Therefore, this study aims to summarize the results of original research articles on the predominant picocyanobacteria in surface water bodies collected from indexed journal articles and gray literature involving the molecular identification of picocyanobacteria. It also provides an understanding of the factors that influence the predominance of picocyanobacteria in these environments, such as trophic status and the method of molecular identification, as well as research trends and the countries that contribute most to this field of research.

2. Materials and methods

2.1 Data collection

This research was conducted as described in the PRISMA Declaration [22]. Thus, for the development of this study, a systematic literature search was carried out in

("Picocyanobacteria" AND lakes OR lagoons AND Freshwater)
("Picocyanobacteria" AND reservoirs OR dams AND freshwater)
("Picocyanobacteria" OR "smallest cyanobacteria" AND freshwater)
("Smallest Cyanobacteria" AND lakes OR lagoons AND freshwater)
("Picocyanobacteria" OR "smallest cyanobacteria" AND lagoons OR lakes AND freshwater)
("Small blue-green algaes" AND "freshwater" OR "sweet water")
("Small blue-green algaes" AND "lakes" OR "lagoons")
("Small blue-green algaes" AND "reservoirs" OR "dams")
("Smallest cyanobacteria" AND "reservoirs" OR "dams")
("Smallest cyanobacteria" AND "freshwater" OR "sweet water")
("Smallest cyanobacteria" AND "lagoons" OR "lakes")
("Picocyanobacteria" AND "reservoirs" OR "dams")

Table 1.
Searches applied for the selection of articles in the three databases used.

three bibliographic databases: Scopus, ScienceDirect, and Scielo, which articles are part of publications in indexed journals [23]; in addition, the Google Scholar search engine was used, focused and specialized in the search for scientific-academic content and bibliography that includes gray literature defined by Garousi et al. [24] as: "literature that is not formally published in sources such as books or journal articles". This allowed an exhaustive search, broadening the information to be analyzed.

Keywords were defined using free terms and the Agrovoc and DeCS thesauri to increase the sensitivity of the search: picocyanobacteria, freshwater, sweetwater, small cyanobacteria, reservoirs, dams, lakes, lagoons, and small blue-green algae. At the same time, Boolean operators (AND and OR) were used to logically connect concepts or groups of terms and to quickly broaden, specify, limit, and define the search (**Table 1**) see annex.

We then proceeded to eliminate duplicate articles using the free tool Zotero-5.0.93. Three investigators independently applied the inclusion and exclusion criteria presented in **Table 2** (see annex) to the resulting articles to avoid bias and ensure reproducibility of the selection.

2.2 Data analysis

The statistical program R Studio® (V 3.6.1) was used to perform the descriptive analysis of the collected data. A database was created using Microsoft Excel where

Inclusion criteria	Exclusion criteria
Original article in english	Brackish water bodies
Molecular identification of picocyanobacteria	Use of collection strains
Field collected sample	

Table 2.
Inclusion and exclusion criteria established and applied to each article for eligibility.

certain attributes of the research were recorded, such as year of publication, the country where the article was published, journal, picocyanobacteria species, water body, trophic state, and picocyanobacteria molecular identification method. Particularly, using this database, the relative and absolute frequencies of the number of publications of picocyanobacteria per year, the number of publications per country, and the number and species of picocyanobacteria most frequently found were determined, as well as some factors related to the prevalence of picocyanobacteria in surface water bodies.

The free software VOSviewer (V 1.6.14) was used to analyze the data on the frequencies of index and author keywords to determine the most frequent keywords researched in the articles included in the systematic review and thus identify trends in research on the topic.

3. Results

3.1 Search strategy and articles obtained

A total of 371 articles were obtained from the databases (Scopus: 251, ScienceDirect: 118, and Scielo: 2) and 57 articles from the Google Scholar search engine. A total of 243 duplicate articles were deleted, resulting in a total of 185 articles subject to eligibility. After applying the inclusion and exclusion criteria, 34 articles were filtered out (**Figure 1**) see annex.

3.2 Articles description

Table 3 shows the detailed information of each of the articles: year of publication, authors, journal in which it was published, the molecular method applied for the identification of picocyanobacteria, and type of water body where the research was carried out.

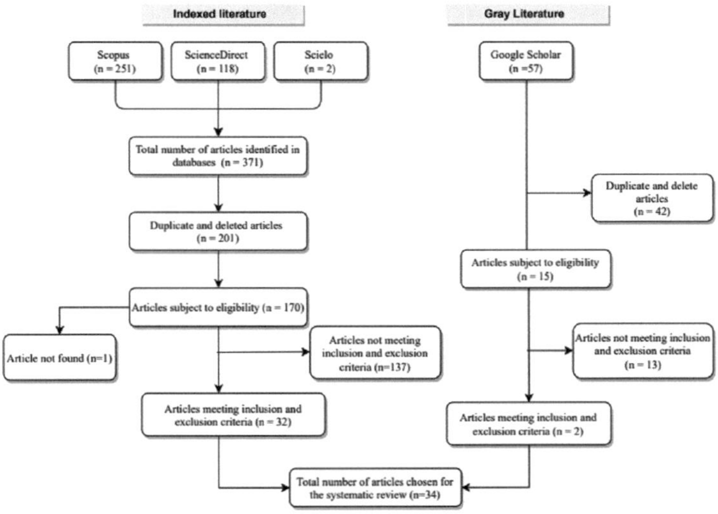

Figure 1.
Flowchart of the research search strategy. Source: own elaboration through the application diagrams.net.

N°	Article name	Year	Author(s)	Journal	Molecular identification method	Water body
1	Sedimentary DNA record of eukaryotic algal and cyanobacterial communities in a shallow Lake driven by human activities and climate change	2021	Hanxiao Zhang, Shouliang Huo, Kevin M. Yeager, Fengchang Wu	Science of The Total Environment	Amplification and Sequencing of the 16S rRNA Gene	Lake
2	Spatiotemporal variability of cyanobacterial community in a Brazilian oligomesotrophic reservoir: The picocyanobacterial dominance	2019	Ana María M. Batista, Alessandra Giani	Ecohydrology & Hydrobiology	Amplification and Sequencing of the 16S rRNA Gene	Reservoir
3	Insights into the evolution of picocyanobacteria and phycoerythrin genes (mpeBA and cpeBA)	2019	Patricia Sánchez Baracaldo, Giorgio Bianchini, Andrea Di Cesare, cristiana Callieri, Nathan A. M. Chrisma	Frontiers in Microbiology	DNA Extraction by PCR and Genome Sequencing	Lake
4	Metabarcoding reveals a more complex cyanobacterial community than morphological identification	2019	Xiao Chuang, LShouliang Huoa, Jingtian Zhanga, Chunzi Ma, Zhe Xiao, Hanxiao Zhang, Beidou Xi, Xinghui Xia	Ecological Indicators	DNA Extraction, Amplification, and Sequencing of the 16S rRNA Gene	Lake and Pond
5	High-throughput DNA sequencing reveals the dominance of pico- and other filamentous cyanobacteria in an urban freshwater Lake	2019	Li, H., Alsanea, A., Barber, M., Goel, R.	Science of the Total Environment	Sequencing of DNA by PCR	Lake
6	Seasonal succession and spatial distribution of bacterial community structure in a eutrophic freshwater Lake, Lake Taihu	2019	Zhu, C., Zhang, J., Nawaz, M.Z., Mahboob, S., Al-Ghanim, K.A., Khan, I.A., Lu, Z., Chen, T	Science of the Total Environment	Amplification and Sequencing of the 16S rRNA Gene	Lake

N°	Article name	Year	Author(s)	Journal	Molecular identification method	Water body
7	Seasonal succession and spatial distribution of bacterial community structure in a eutrophic freshwater Lake, Lake Taihu	2018	Zhu, C., Zhang, J., Nawaz, M.Z., Mahboob, S., Al-Ghanim, K.A., Khan, I.A., Lu, Z., Chen, T	Science of the Total Environment	Amplification and Sequencing of the 16S rRNA Gene	Lake
8	Ecological and genomic features of two widespread freshwater picocyanobacteria	2018	Cabello-Yeves, P.J., Picazo, A., Camacho, A., Callieri, C., Rosselli, R., Roda-Garcia, J.J., Coutinho, F.H., Rodriguez-Valera, F.	Environmental Microbiology	Sequencing of DNA by PCR	Reservoir
9	Planktonic cyanobacteria from a tropical reservoir of Southeastern Brazil: A picocyanobacteria rich community and new approaches for its characterization	2018	Marcele Laux, Vera Regina Werner, Ricardo A. Vialle, José Miguel Ortega, Alessandra Giani	Nova Hedwigia	Amplification of DNA by PCR	Reservoir
10	Novel *Synechococcus* genomes reconstructed from freshwater reservoirs	2017	Cabello-Yeves, P.J., Haro-Moreno, J.M., Martin-Cuadrado, A.-B., Ghai, R., Picazo, A., Camacho, A., Rodriguez-Valera, F.	Frontiers in Microbiology	Amplification and Sequencing of the 16S rRNA Gene	Reservoir
11	Metagenomic analysis in Lake Onego (Russia) *Synechococcus* cyanobacteria	2017	Vasileva, A., Skopina, M., Averina, S., Gavrilova, O., Ivanikova, N., Pinevich, A	Journal of Great Lakes Research	Amplification and Sequencing of the 16S rRNA Gene	Lake
12	Phenotypic plasticity in freshwater picocyanobacteria	2017	Huber, P., Diovisalvi, N., Ferraro, M., Metz, S., Lagomarsino, L., Llames, M.E., Royo-Llonch, M., Bustingorry, J., Escaray, R.	Environmental Microbiology	Amplification and Sequencing of the 16S rRNA Gene	Lake

N°	Article name	Year	Author(s)	Journal	Molecular identification method	Water body
13	Microbial community structure and interannual change in the last epishelf lake ecosystem in the north polar region	2017	Taller, M., Vincent, W.F., Lionard, M., Hamilton, A.K., Lovejoy, C	Frontiers in Marine Science	Amplification and Sequencing of the 16S rRNA Gene	Lake
14	*Synechococcus* diversity along a trophic gradient in the Osterseen Lake District, Bavaria	2016	Ruber, J., Bauer, F.R., Millard, A.D., Raeder, U., Geist, J., Zwirglmaier, K	Microbiology (United Kingdom)	Amplification and Sequencing of the 16S rRNA Gene	Lake
15	CO2 alters picophytoplankton community structure in freshwater ecosystems	2016	Shi, X., Li, S., Wang, X., Liu, M., Kong, F.	Fundamental and Applied Limnology	DNA Sequencing of 18S RNA Genes	Lake
16	Community analysis of picocyanobacteria in an oligotrophic lake by cloning 16S rRNA gene and 16S rRNA gene amplicon sequencing	2015	Fujimoto,N., Mizuno, K., Yokoyama, T., Ohnishi, A., Suzuki, M., Watanabe, S., Komatsu, K., Sakata, Y., Kishida, N., Akiba, M., Matsukura, S.	Journal of General and Applied Microbiology	Amplification and Sequencing of the 16S rRNA Gene	Lake
17	Diversity of Lake Ladoga (Russia) bacterial plankton inferred from 16S rRNA gene pyrosequencing: An emphasis on picocyanobacteria	2015	Skopina, M., Pershina, E., Andronov, E., Vasileva, A., Averina, S., Gavrilova, O., Ivanikova, N., Pinevich, A.	Journal of Great Lakes Research	Amplification and Sequencing of the 16S rRNA Gene	Lake
18	Genetic diversity of picocyanobacteria in Tibetan lakes: Assessing the endemic and universal distributions	2014	Huang, S., Liu, Y., Hu, A., Liu, X., Chen, F., Yao, T., Jiao, N.	Applied and Environmental Microbiology	Amplification and Sequencing of the 16S and 23S rRNA Genes	Lake
19	Free-living and particle-associated bacterioplankton in large rivers of the Mississippi River basin demonstrate biogeographic patterns	2014	Colin R. Jackson, Justin J. Millar, Jason T. Payne, Clifford A. Ochs	Applied and Environmental Microbiology	DNA Extraction, Amplification, and Sequencing of the 16S rRNA Gene	River

N°	Article name	Year	Author(s)	Journal	Molecular identification method	Water body
20	Detection and expression of genes for phosphorus metabolism in picocyanobacteria from the Laurentian Great Lakes	2013	Kutovaya, O.A., McKay, R.M.L., Bullerjahn, G.S.	Journal of Great Lakes Research	Sequencing of DNA by PCR	Lake
21	Seasonal and Spatial Diversity of Picocyanobacteria Community in the Great Mazurian Lakes Derived from DGGE Analyses of 16S rDNA and cpcBA-IGS Markers	2013	Jasser, I., Królicka, A., Jakubiec, K., Chróst, RJ	Journal of Microbiology and Biotechnology	DGGE analysis of molecular markers derived from the 16S–23S internal transcribed spacer (ITS) of the ribosomal operon.	Lake
22	Picocyanobacterial community structure and space–time dynamics in the subalpine Lake Maggiore (N. Italy)	2012	Callieri, C., Caravati, E., Corno, G., Bertoni, R.	Journal of Limnology	Amplification and Sequencing of the 16S and 23S rRNA Genes	Lake
23	Genome sequences of siphoviruses infecting marine *Synechococcus* unveil a diverse cyanophage group and extensive phage-host genetic exchanges	2012	Sijun Huang, Kui Wang, Nianzhi Jiao, Feng Chen	Environmental Microbiology	Amplification and Sequencing of the 16S rRNA Gene	Bay
24	Vertical and longitudinal distribution patterns of different bacterioplankton populations in a canyon-shaped, deep prealpine lake	2011	Salcher, M.M., Pernthaler, J., Frater, N., Posch, T.	Limnology and Oceanography	Amplification and Sequencing of the 16S rRNA Gene	Lake
25	East Tibetan lakes harbor novel clusters of picocyanobacteria as inferred from the 16S–23S rRNA internal transcribed spacer sequences	2010	Wu, Q.L., Xing, P., Liu, W.-T.	Microbial Ecology	Fragment Polymorphism Analysis of 16S–23S rRNA Internal Transcribed Spacer (ITS) PCR Amplicon	Lake
26	Photosynthetic picoplankton dynamics in Lake Tahoe: Temporal and spatial niche partitioning among prokaryotic and eukaryotic cells	2009	Winder, M.	Journal of Plankton Research	Phycoerythrin (PE) and Chlorophyll (Chl) Fluorescence by Cytogram	Lake

N°	Article name	Year	Author(s)	Journal	Molecular identification method	Water body
27	High ratio of bacteriochlorophyll biosynthesis genes to chlorophyll biosynthesis genes in bacteria of humic lakes	2009	Eiler, A., Beier, S., Säwström, C., Karlsson, J., Bertilsson, S.	Applied and Environmental Microbiology	Sequencing of DNA by PCR	Lake
28	Lake superior supports novel clusters of cyanobacterial picoplankton	2007	Ivanikova, N.V., Popels, L.C., McKay, R.M.L., Bullerjahn, G.S.	Applied and Environmental Microbiology	Sequencing of 16S rRNA Gene and cpcBA Phycocyanin Operon Intergenic Spacer (IGS) Sequences	Lake
29	Photosynthetic characteristics and diversity of freshwater Synechococcus at two depths during different mixing conditions in a deep oligotrophic lake	2007	Callieri, C., Corno, G., Caravati, E., Galafassi, S., Bottinelli, M., Bertoni, R.	Journal of Limnology	Amplification and Sequencing of the 16S rRNA Gene	Lake
30	Abundance and diversity of picocyanobacteria in High Arctic lakes and fjords	2006	Patrick Van Hove, Warwick F. Vincent, Pierre E. Galand, Annick Wilmotte	Algological studies	DNA Extraction, Amplification, and Sequencing of the 16S rRNA Gene	Lake
31	Rapid establishment of clonal isolates of freshwater autotrophic picoplankton by single-cell and single-colony sorting	2003	Crosbie, N.D., Pöckl, M., Weisse, T.	Journal of Microbiological Methods	Direct Sequencing of the 16S rRNA Gene and cpcBA-IGS Region	Lake
32	Dispersal and phylogenetic diversity of nonmarine picocyanobacteria, inferred from 16S rRNA gene and cpcBA-intergenic spacer sequence analyses	2003	Crosbie, N.D., Pöckl, M., Weisse, T.	Applied and Environmental Microbiology	Amplification and Sequencing of the 16S rRNA Gene	Lake
33	Identification of cultured and uncultured picocyanobacteria from a mesotrophic freshwater lake based on the partial sequences of 16S rDNA	2001	Toshiya Katano Manabu Fukui Yasunori Watanabe	Limnology	Amplification and Sequencing of the 16S rRNA Gene	Lake

N°	Article name	Year	Author(s)	Journal	Molecular identification method	Water body
34	Systematics and ecology of chlorophyte picoplankton in German Inland waters along a nutrient gradient	2001	Dominik Hepperle., Lothar Krienitz	International Review of Hydrobiology	Amplification of DNA by PCR	Lake

Table 3.
Summary of the results of each article selected for the research.

The number of publications per year ranged from 2001 to 2021. Of the 34 articles analyzed, it was found that 2019 was the year with the highest number of publications recorded on the subject, followed by 2017, which had five and four publications, respectively (**Table 3**) see annex. This is evidence that research on picocyanobacteria has increased in recent years.

In addition, **Figure 2** (see annex) shows the countries with the greatest number of research projects developed for the study and identification of picocyanobacteria in the environments described above. Thus, the country where the institute or center where the research was carried out is located was identified. It is important to clarify that these countries corresponded to the sampling sites.

The countries with the highest number of research studies were China (7), United States (6), and United Kingdom (3). The countries with at least two publications were: Brazil, Canada, Spain, Italy, Japan, and Russia. Germany, Argentina, Switzerland, Sweden, Austria, and Poland had only one publication.

Figure 3 (see annex) shows the picocyanobacteria identified in the articles analyzed, showing that *Synechococcus* was the predominant genus in surface water bodies, with a frequency of 24 articles, followed by *Cyanobium* with a frequency of 4 articles.

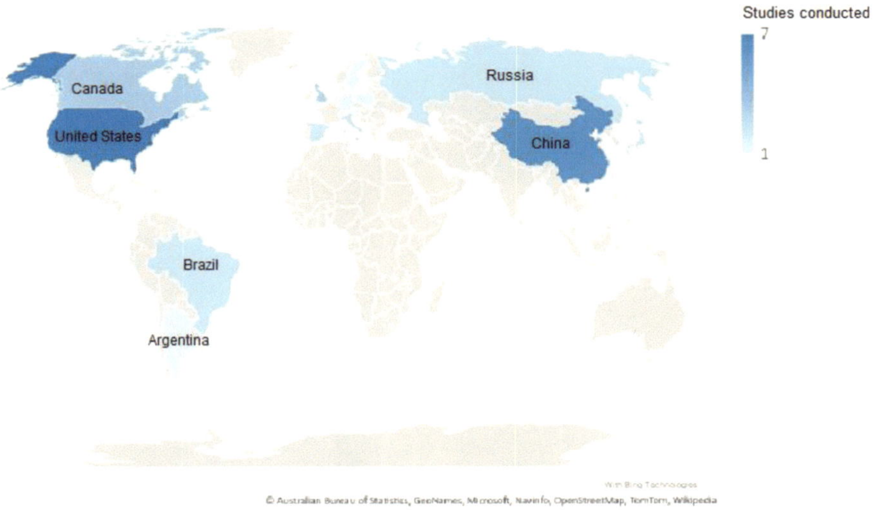

Figure 2.
Countries where research studies on picocyanobacteria molecularly identified in surface waters have been carried out.Source: own elaboration through the software program Microsoft Excel.

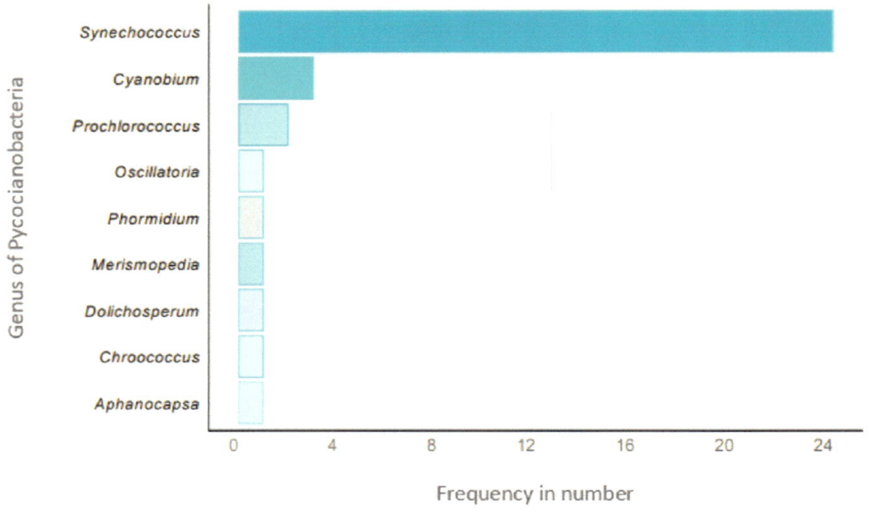

Figure 3.
Predominant picocyanobacteria in freshwater bodies. Source: own elaboration through the statistical tool RStudio.

As shown in **Figure 4** (see annex), oligotrophic lakes were the most studied for the identification of picocyanobacteria with a relative frequency of 32%, followed by oligomesotrophic lakes with 18.6% and mesotrophic and eutrophic lakes with 14% each.

3.3 Research topics on picocyanobacteria

The keyword mapping shows that the words: Cyanobacterium, Cyanobacteria, *Synechococcus*, Lake, Microbiology, 16S RNA, and Picocyanobacteria are the ones that

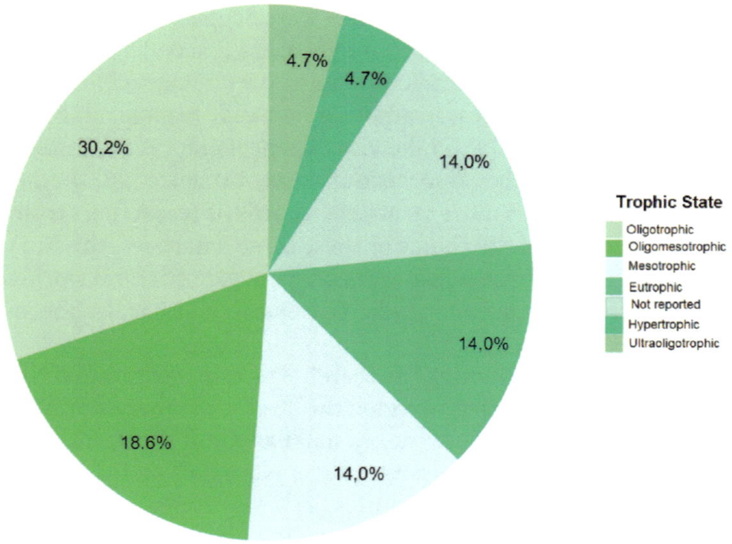

Figure 4.
Trophic state of the lakes studied in the articles of interest. Source: own elaboration through the statistical tool RStudio.

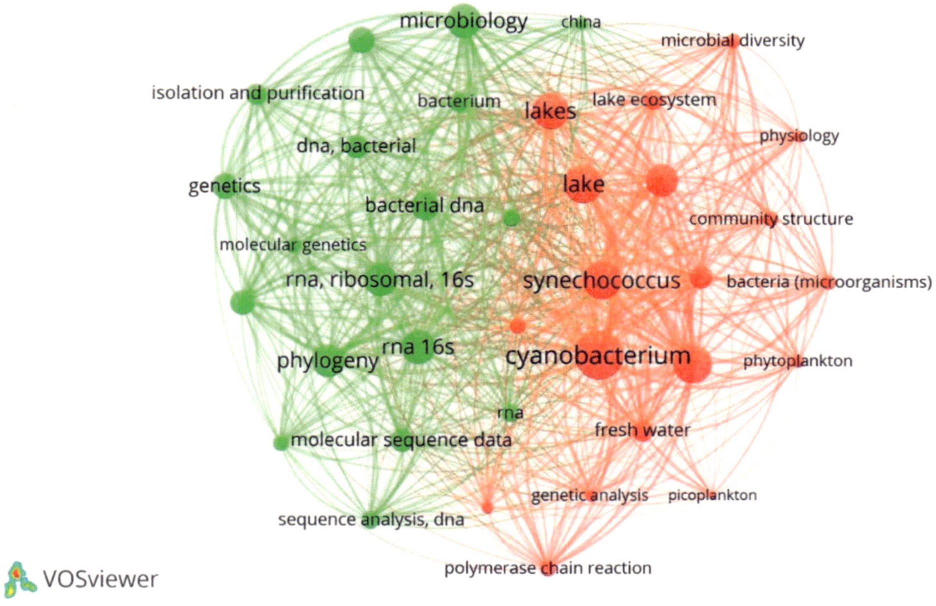

Figure 5.
Keyword mapping used in the search. Source: own elaboration through the application VOSviewer.

show the highest tendency in the present research with a frequency in the number of articles of 22, 18, 16, 14, 11, 11 and 10, respectively (**Figure 5**) see annex.

4. Discussion

The databases used in this systematic review were Scopus, Science Direct, and Scielo, as mentioned above. Scopus was chosen because it is widely known as one of the largest databases of abstracts and citations of peer-reviewed literature and has many records in the science area. In addition, it is easy to export bibliographic information for further analysis [25]. Likewise, ScienceDirect is a database with an extensive record of article records in various areas of science [26]. In the case of Scopus, this database focuses more on article records of researchers from South American countries [27]. With the choice of these three databases, the aim was to address the largest number of research studies on picocyanobacteria worldwide, since this is a subject that has not been studied extensively, as has been recognized by several authors.

In this regard, and after conducting the search strategy, it was found that the ScienceDirect and Scopus databases provided the largest number of publications due to their worldwide positioning as indexed databases and their mainly English-language journals. In addition, Scopus covers various areas of science, technology, medicine, social sciences, arts, and humanities [25]. Moreover, Sciencedirect is a database that also covers multidisciplinary scientific areas [28], however, it is limited to journals and books published directly by its publisher [29]. Therefore, although many of the articles included in this review were found, they did not exceed those found in Scopus [30].

In contrast, the Scielo open-access database, although it includes journals from all areas of science, only two articles associated with the topic were found in the search. This is since this database contains scientific articles published only in Latin America, and because it is a database that publishes mainly in Spanish and Portuguese [31].

There are different molecular techniques used for the identification of picocyanobacteria, in this review we found that the application of these techniques to characterize and amplify portions of the cyanobacterial genome has increased considerably in recent years. These techniques have proven to be valuable for comparing the structures of complex microbial communities, inferring phylogenetic relationships, and monitoring their dynamics in relation to environmental factors [32]. Cyanobacteria such as *Synechococcus* and *Cyanoothece* are particularly difficult to identify and classify [33], most molecular methods to identify them are based on total DNA or RNA extraction and amplification by PCR as shown in **Table 3**. However, there are biases related to the presence of PCR inhibitors and primer specificity and efficiency that can skew the results of community composition [33].

Concerning the number of annual publications obtained in the analyzed period (**Table 3**), there was consistency with the findings of Rousso et al. [4] in their research on predictive models for cyanobacterial blooms in freshwater lakes. They found that in the period between 2014 and 2019 the highest number of publications on cyanobacteria was reported, the same as this research, where it was found that between 2015 and 2019 the highest number of publications on picocyanobacteria was collected. However, the maximum number of publications found for the articles found that met the inclusion criteria was only 5 for the year 2019, indicating that there are still not many studies on picocyanobacteria [34]. Research on cyanobacteria appears to be strongly related to advances in monitoring technology, i.e., increased availability of data, knowledge of cyanobacterial ecology, physiology, and risks, among other factors [34]. Furthermore, due to the environmental problems associated with cyanobacteria and their potentially toxic blooms. Merel et al. [8] evidence that articles on cyanobacteria have increased significantly in the period 1995–2010, a trend that is expected to continue [8]. This systematic review shows that 2010 was a year where no significant reports on picocyanobacteria were found, which probably indicates that research is still focused on microplankton cyanobacteria instead of picocyanobacteria.

Regarding the countries where the studies were conducted, it was found that the United States and China have been outstanding countries for the number of scientific publications on cyanobacteria and their toxic blooms, this is demonstrated by the study conducted by Bertone [4], which analyzed the publications on CyanoHAB in different countries and found that most of the publications are focused on the United States, Northern Europe, Southeast China, Japan, and Oceania [4]. This result coincides with that found in this study (**Figure 2**), where the highest number of scientific articles on picocyanobacteria have been published in the United States and China.

The concentration of publications in developed countries such as these may be related to their economy and extensive scientific resources, the provision of funds for research and development, and the availability of data used for these purposes [35]. Also, these countries have managed to develop specialized monitoring and control procedures for CyanoHAB from research [8, 36]. On the other hand, Ndlela [37], made an overview of cyanobacterial bloom occurrences and research in Africa during the last decade and found that the amount of information available on the continent on the subject is limited probably due to the general inadequacy of the infrastructure and its relation to civil wars [37].

Regarding the most frequent genera, the genus *Synechococcus* was the most reported with a frequency of 24; this genus plays a fundamental role in the ecology of surface water bodies that are important human resources, being predominant in freshwater systems. Generally, picocyanobacteria of the genus *Synechococcus*, *Prochlorococcus,* and *Cyanobium* are designated as non-flourishing [38]. However, some strains of the genus *Synechococcus* can produce toxins such as β-N-methylamino-L-alanine (BMAA), and microcystin (MC) [39], which causes problems in the ecosystem and human health. Similarly, Gin [15] through his study showed that *Synechococcus* spp. could produce cylindrospermopsin (CYN) and anatoxin-a (ATX) which are alkaloids that can cause damage to mammalian organisms, this discovery has implications on the potential risk to freshwater resources that serve as drinking water supply [40].

Previous studies by Li [21] report that the prevalence of *Synechococcus* in water bodies is influenced by warm temperature, high nutrient level, and phosphorus limitation, comprising fractions of up to 80% of the total biomass of picocyanobacteria of a bloom [41]. Furthermore, the result obtained in this review agrees with that reported by Cabello [38], where it is confirmed that the genera *Synechococcus* and *Cyanobium* are the dominant picocyanobacteria in freshwater systems [42].

Prochlorococcus ranks third as one of the most frequently found picocyanobacteria in research. It inhabits the entire photic zone and can be found as deep as 200 m below the surface, being abundant in oligotrophic systems [43]. *Prochlorococcus* and *Synechococcus* can coexist in water bodies, but *Synechococcus* tolerates a wider temperature range, without being limited by temperatures as low as 2°C and is more ubiquitous and has a wider latitudinal distribution [32].

It has been shown that the trophic state of water bodies influences the composition and abundance of picocyanobacterial communities. It was observed that the most studied lakes were those in an oligotrophic state, these lakes are characterized by being poor in nutrients and having low primary productivity [44], which limits the presence of a high microbial density and only those taxa that have adapted to these conditions can survive. Thus, picocyanobacteria of the genus *Synechococcus* are predominant in these systems, this is due to the ability of these picocyanobacteria to adapt to low light conditions, their affinity for orthophosphate and other sources of inorganic phosphorus, as well as their ability to store nitrogen in phycobilins that increase the competition of *Synechococcus* against algae and other bacteria, as stated by Vanstein [45]. The above is consistent with that reported by Joachim Ruber et al. [46], who describe this important genus as dominant in oligotrophic conditions, concluding that *Synechococcus* could be used as a bioindicator in such environments [46]. Besides, authors such as Rousso et al. [4] reported in a systematic review on CyanoHAB that more than 50% of the lakes investigated were eutrophic or hypertrophic and only 8% of the lakes were oligotrophic, reporting that the occurrence of CyanoHAB is related to the levels of nutrients present in the lakes [47].

In **Figure 5**, it is evident that the distance between two keywords demonstrates relative strength and similarity of topic and circles in the same color group suggest that a similar topic is being addressed among the publications [48]. **Figure 5** shows that the most used keywords are: *Synechococcus*, microbial community, phylogeny, and 16S rRNA, this shows us that more molecular identification strategies have been used recently for the identification of picocyanobacteria as cited by Demoulin et al. [49], who indicate that since the late 1990's many phylogenetic studies based on 16S rRNA or specific protein have been published. Similarly, the results of the keywords are also observed in **Figure 5**, in which the most frequent words have been used a greater number of times in the articles. The total link strength attribute indicates when a

keyword is very important because it is identified to have had a lot of interaction with other keywords in the analyzed articles, the higher the value the stronger the link that exists between one word and another [50].

The findings of the current systematic review show the lack of research on picocyanobacteria in surface waters that allow understanding the importance they represent as beneficial microorganisms; standing out for being part of the primary producers, or harmful because they can produce toxic blooms. It was also evidenced that molecular identification methods of picocyanobacteria have recently begun to be highlighted in research methodologies, which shows a transition from traditional research to a more advanced one.

5. Conclusions

Although in the last two decades the identification of picocyanobacteria has increased due to the implementation of new automated methods and molecular techniques, studies aimed at identifying them in surface water bodies intended for recreational use or drinking water supply are still incipient, which is possibly explained by the difficulty in their characterization and rapid physiological plasticity. The predominant genus of picocyanobacteria in this systematic review was *Synechococcus*, a producer of toxic compounds, which generates an alert and highlights the importance of advancing in the implementation of protocols for sampling and identification of these bacteria for epidemiological surveillance.

The countries where more studies on cyanobacteria were conducted were the United States and China since these are developed countries that invest their resources in education and research and can develop specialized monitoring and control procedures for CyanoHAB from their scientific resources. Therefore, there is a need for further research in this area, to use the information for further studies and decision making.

Acknowledgements

The authors would like to thank the support provided by the Research Group on Health and Sustainability of the University of Antioquia, as well as the School of Microbiology who, through the financing of projects for degree work, made possible the realization and completion of this project.

Conflicts of interest

There are no conflicts of interest disclosed for this publication.

Consent to publication

The authors involved gave their consent for publication.

Ethical approval and consent to participate

The study does not involve data related to animal or human participation.

Author details

Alejandra Sandoval Valencia, Lisseth Dahiana Salas, María Alejandra Pérez Gutiérrez, Luisa María Munera Porras* and Leonardo Alberto Ríos-Osorio
Research Group on Health and Sustainability, School of Microbiology, University of Antioquia, Medellin, Colombia

*Address all correspondence to: luisam.munera@udea.edu.co

References

[1] Cantoral E. Cianotoxinas: Efectos ambientales y sanitarios. Medidas de prevención. 2017. Available from: http://www.scielo.org.mx/scielo.php?script=sci_arttext&pid=S0188-88972017000200241

[2] Liu L, Huang Q, Qin B. Characteristics and roles of Microcystis extracellular polymeric substances (EPS) in cyanobacterial blooms: A short review. Journal of Freshwater Ecology. 2018;**33**(1):183-193. DOI: 10.1080/02705060.2017.1391722

[3] Huisman J, Codd GA, Paerl HW, Ibelings BW, Verspagen JMH, Visser PM. Cyanobacterial blooms. Nature Reviews Microbiology. 2018;**16**(8):471-483. DOI: 10.1038/s41579-018-0040-1

[4] Rousso BZ, Bertone E, Stewart R, Hamilton DP. A systematic literature review of forecasting and predictive models for cyanobacteria blooms in freshwater lakes. Water Research. 2020. DOI: 10.1016/j.watres.2020.115959

[5] Loftin KA, Graham JL, Hilborn ED, Lehmann SC, Meyer MT, Dietze JE, et al. Cyanotoxins in inland lakes of the United States: Occurrence and potential recreational health risks in the EPA National Lakes Assessment 2007. Harmful Algae. 2016;**56**:77-90. DOI: 10.1016/j.hal.2016.04.001

[6] Qin B, Li W, Zhu G, Zhang Y, Wu T, Gao G. Cyanobacterial bloom management through integrated monitoring and forecasting in large shallow eutrophic Lake Taihu (China). Journal of Hazardous Materials. 2015;**287**:356-363. DOI: 10.1016/j.jhazmat.2015.01.047

[7] Suurnäkki S, Gomez-Saez GV, Rantala-Ylinen A, Jokela J, Fewer DP, Sivonen K. Identification of geosmin and 2-methylisoborneol in cyanobacteria and molecular detection methods for the producers of these compounds. Water Research. 2015;**68**:56-66. DOI: 10.1016/j.watres.2014.09.037

[8] Merel S, Walker D, Chicana R, Snyder S, Baurès E, Thomas O. State of knowledge and concerns on cyanobacterial blooms and cyanotoxins. Environment International. 2013;**59**:303-327

[9] Zamyadi A, Coral LA, Barbeau B, Dorner S, Lapolli FR, Prévost M. Fate of toxic cyanobacterial genera from natural bloom events during ozonation. Water Research. 2015;**73**:204-215. DOI: 10.1016/j.watres.2015.01.029

[10] Kormas KA, Gkelis S, Vardaka E, Moustaka-Gouni M. Morphological and molecular analysis of bloom-forming Cyanobacteria in two eutrophic, shallow Mediterranean lakes. Limnologica. 2011;**41**(3):167-173. DOI: 10.1016/j.limno.2010.10.003

[11] Vardaka E, Moustaka-Gouni M, Cook CM, Lanaras T. Cyanobacterial blooms and water quality in Greek waterbodies. Journal of Applied Phycology. 2005;**17**(5):391-401. DOI: 10.1007/s10811-005-8700-8

[12] Jakubowska N, Szeląg-Wasielewska E. Toxic picoplanktonic cyanobacteria—Review. Marine Drugs. 2015;**13**(3):1497-1518. DOI: 10.3390/md13031497

[13] Śliwińska-Wilczewska S, Maculewicz J, Felpeto AB, Latała A. Allelopathic and bloom-forming picocyanobacteria in a changing world. Toxins. 2018;**10**(1):48. DOI: 10.3390/toxins10010048

[14] Jasser I, Callieri C. Picocyanobacteria. In: Meriluoto J, Spoof L, Codd GA, editors. Handbook of Cyanobacterial Monitoring and Cyanotoxin Analysis. 2017. pp. 19-27. DOI: 10.1002/9781119068761.ch3

[15] Gin KYH, Sim ZY, Goh KC, Kok JWK, Te SH, Tran NH, et al. Novel cyanotoxin-producing Synechococcus in tropical lakes. Water Research. 2021;**192**:116828. DOI: 10.1016/j.watres.2021.116828

[16] Sánchez-Baracaldo P, Bianchini G, Di Cesare A, Callieri C, Chrismas NAM. Insights into the evolution of picocyanobacteria and phycoerythrin genes (mpeBA and cpeBA). Frontiers in Microbiology. 2019;**10**(JAN):45. DOI: 10.3389/fmicb.2019.00045

[17] Álvarez E, Moyano M, López-Urrutia Á, Nogueira E, Scharek R. Routine determination of plankton community composition and size structure: A comparison between FlowCAM and light microscopy. Journal of Plankton Research. 2014;**36**(1):170-184. DOI: 10.1093/plankt/fbt069

[18] Tamulonis C, Kaandorp J. A model of filamentous cyanobacteria leading to reticulate pattern formation. Life. 2014;**4**(3):433-456. DOI: 10.3390/life4030433

[19] Gkelis S, Papadimitriou T, Zaoutsos N, Leonardos I. Anthropogenic and climate-induced change favors toxic cyanobacteria blooms: Evidence from monitoring a highly eutrophic, urban Mediterranean lake. Harmful Algae. 2014;**39**:322-333. DOI: 10.1016/j.hal.2014.09.002

[20] Bury-Moné S, Sclavi B. Stochasticity of gene expression as a motor of epigenetics in bacteria: from individual to collective behaviors. Research in Microbiology. 2017;**168**(6):503-514. DOI: 10.1016/j.resmic.2017.03.009

[21] Li J, Chen Z, Jing Z, Zhou L, Li G, Ke Z, et al. Synechococcus bloom in the Pearl River Estuary and adjacent coastal area–With special focus on flooding during wet seasons. Science of the Total Environment. 2019;**692**:769-783. DOI: 10.1016/j.scitotenv.2019.07.088

[22] Urrutia G, Bonfill X. Declaración PRISMA: una propuesta para mejorar la publicación de revisiones sistemáticas y metaanálisis. Medicina Clínica. 2010:507-511. Available from: http://www.laalamedilla.org/Investigacion/Recursos/PRISMA Spanish Sept 2010.pdf

[23] Moed HF, De Moya-Anegon F, Guerrero-Bote V, Lopez-Illescas C. Are nationally oriented journals indexed in Scopus becoming more international? The effect of publication language and access modality. Journal of Informetrics. 2020;**14**(2):101011. DOI: 10.1016/j.joi.2020.101011

[24] Garousi V, Felderer M, Mäntylä MV. Guidelines for including grey literature and conducting multivocal literature reviews in software engineering. Information and Software Technology. 2019;**106**:101-121. DOI: 10.1016/j.infsof.2018.09.006

[25] About Scopus-Abstract and citation database. Elsevier; n.d.. Available from: https://www.elsevier.com/solutions/scopus. [Accessed: May 23, 2021]

[26] Valencia C. Science Direct. 2021. Available from: https://biblioguias.webs.upv.es/bg/index.php/es/libros-electronicos/science-direct

[27] Martín-Martín A, Orduna-Malea E, Thelwall M, Delgado López-Cózar E. Google Scholar, Web of Science, and Scopus: A systematic comparison of citations in 252 subject categories. Journal of Informetrics. 2018;**12**(4):1160-1177. DOI: 10.1016/j.joi.2018.09.002

[28] Bravo Á. Comparativo de dos bases de datos de literatura científica. 2013. Available from: http://www.scielo.org.mx/scielo.php?script=sci_arttext&pid=S0187-73802013000200001

[29] Codina L. Science Direct: Base de datos y plataforma digital de Elsevier. 2018. Available from: https://www.lluiscodina.com/science-direct-elsevier/

[30] Álvarez de Toledo ML. Infobiblio. Información Bibliográfica: Scopus y Science Direct. Qué utilizar; 2011. Availanle from: http://infobib.blogspot.com/2010/01/scopus-y-science-direct-que-utilizar.html

[31] Bojo Canales CMFC. SciELO: Un proyecto cooperativo para la difusión de la ciencia. 2009. Availanle from: https://scielo.isciii.es/scielo.php?script=sci_arttext&pid=S1575-06202009000200004

[32] Santana-Vega Z, Hernández-Becerril DU, Morales-Blake AR, Varona-Cordero F, Merino-Ibarra M. Prokaryotic picoplankton distribution within the oxygen minimum zone of the central mexican pacific across environmental gradients. Brazilian Journal of Oceanography. 2018;**66**(2):157-171. DOI: 10.1590/s1679-87592018004806602

[33] Gupta V, Ratha SK, Sood A, Chaudhary V, Prasanna R. New insights into the biodiversity and applications of cyanobacteria (blue-green algae)—Prospects and challenges. Algal Research. 2013;**2**(2):79-97. DOI: 10.1016/j.algal.2013.01.006

[34] Zamyadi A, Choo F, Newcombe G, Stuetz R, Henderson RK. A review of monitoring technologies for real-time management of cyanobacteria: Recent advances and future direction. TrAC-Trends in Analytical Chemistry. 2016;**85**:83-96. DOI: 10.1016/j.trac.2016.06.023

[35] Cao H, Recknagel F, Orr PT. Enhanced functionality of the redesigned hybrid evolutionary algorithm HEA demonstrated by predictive modelling of algal growth in the Wivenhoe Reservoir, Queensland (Australia). Ecological Modelling. 2013;**252**(1):32-43. DOI: 10.1016/j.ecolmodel.2012.09.009

[36] Carmichael W. A world overview—One-hundred-twenty-seven years of research on toxic cyanobacteria—Where do we go from here? In: Advances in Experimental Medicine and Biology. Vol. 619. New York: Springer; 2008. pp. 104-125. DOI: 10.1007/978-0-387-75865-7_4

[37] Ndlela LL, Oberholster PJ, Van Wyk JH, Cheng PH. An overview of cyanobacterial bloom occurrences and research in Africa over the last decade. Harmful Algae. 2016;**60**:11-26. DOI: 10.1016/j.hal.2016.10.001

[38] Cabello-Yeves PJ, Haro-Moreno JM, Martin-Cuadrado A-B, Ghai R, Picazo A, Camacho A, et al. Novel synechococcus genomes reconstructed from freshwater reservoirs. Frontiers in Microbiology. 2017;**8**(Jun):1151. DOI: 10.3389/fmicb.2017.01151

[39] Adamski M, Wołowski K, Kaminski A, Hindáková A. Cyanotoxin cylindrospermopsin producers and the catalytic decomposition process: A review. Harmful Algae. 2020;**98**:101894. DOI: 10.1016/j.hal.2020.101894

[40] Lin SS, Shen SL, Zhou A, Lyu HM. Assessment and management of lake eutrophication: A case study in Lake Erhai, China. Science of the Total Environment. 2021;**751**:141618. DOI: 10.1016/j.scitotenv.2020.141618

[41] Śliwińska-WilczewskaS,MaculewiczJ, Tuszer J, Dobosz K, Kulasa D, Latała A. First record of allelopathic activity of the picocyanobacterium Synechococcus

sp. on a natural plankton community. Ecohydrology and Hydrobiology. 2017;**17**(3):227-234. DOI: 10.1016/j.ecohyd.2017.05.001

[42] Callieri C. Picophytoplankton in freshwater ecosystems: The importance of small-sized phototrophs. Freshwater Reviews. 2008;**1**(1):1-28. DOI: 10.1608/FRJ-1.1.1

[43] Al-Hosani S, Oudah MM, Henschel A, Yousef LF. Global transcriptome analysis of salt acclimated Prochlorococcus AS9601. Microbiological Research. 2015;**176**:21-28. DOI: 10.1016/j.micres.2015.04.006

[44] Li H, Alsanea A, Barber M, Goel R. High-throughput DNA sequencing reveals the dominance of pico- and other filamentous cyanobacteria in an urban freshwater Lake. Science of the Total Environment. 2019;**661**:465-480. DOI: 10.1016/j.scitotenv.2019.01.141

[45] Vadstein O. Heterotrophic, planktonic bacteria and cycling of phosphorus phosphorus requirements, competitive ability, and food web interactions. Advances in Microbial Ecology. 2000;**16**(1):115-167. DOI: 10.1007/978-1-4615-4187-5_4

[46] Ruber J, Bauer FR, Millard AD, Raeder U, Geist J, Zwirglmaier K. Synechococcus diversity along a trophic gradient in the Osterseen Lake District, Bavaria. Microbiology (United Kingdom). 2016;**162**(12):2053-2063. DOI: 10.1099/mic.0.000389

[47] Paerl HW, Hall NS, Calandrino ES. Controlling harmful cyanobacterial blooms in a world experiencing anthropogenic and climatic-induced change. Science of the Total Environment. 2011;**409**(10):1739-1745. DOI: 10.1016/j.scitotenv.2011.02.001

[48] Guo Y-M, Huang Z-L, Guo J, Li H, Guo X, Nkeli M. Bibliometric analysis on smart cities research. Sustainability. 2019;**11**:3606. DOI: 10.3390/su11133606

[49] Demoulin CF, Lara YJ, Cornet L, François C, Baurain D, Wilmotte A, et al. Cyanobacteria evolution: Insight from the fossil record. Free Radical Biology and Medicine. 2019;**140**:206-223. DOI: 10.1016/j.freeradbiomed.2019.05.007

[50] Jan van Eck N, Waltman L. VOSviewer Manual. 2018. Available from: https://www.vosviewer.com/documentation/Manual_VOSviewer_1.6.8.pdf

Chapter 2

Removal of Microcystins from Drinking Water by Electrocoagulation: Upscaling, Challenges, and Prospects

Stephen Opoku-Duah, Dennis Johnson, Dan Blair and Jeff Dimick

Abstract

Microcystins (MCs) belong to a family of stable monocyclic heptapeptide compounds responsible for hazardous toxins in drinking water. Although several methods have been applied to remove MCs from drinking water (e.g., activated carbon filtration, ion exchange resins, high-pressure membranes, and electrochemistry), upscaling laboratory experiments to benefit municipal water treatment is still a major challenge. This chapter is a follow-up study designed to test three electrocoagulation (EC) techniques for decomposing MC by UV-ozone purification (laboratory), electrocoagulation (field unit), and coupled UV-ozone-electrocoagulation (municipal treatment). The chemistry and efficiency of the treatments were first examined followed by comparison with activated carbon filtration. Electrocoagulation outperformed activated carbon filtration by nearly 40%. When the laboratory treatments were evaluated at the municipal scale, effectiveness of the technique deteriorated by 10–20% because of UV pulse dissipation, vapor-ion plasma underfunctioning, and limitations of polymer fiber filters. We confirmed previously published studies that pollutant coagulation and MC decomposition are affected by physicochemical factors such as radiation pulse density, electrical polarity, pH, and temperature dynamics. The results have relevant applications in wastewater treatment and chemical recycling.

Keywords: microcystins, drinking water, UV-ozone purification, electrocoagulation, municipal, coupled UV-electrocoagulation

1. Introduction

Cyanobacteria (also called cyanotoxins) in drinking water is a global concern because of their hazardous effects on human and animal health [1–3]. Microcystins (MCs) are a common source of cyanotoxins. MCs are produced by a variety of cyanobacteria including *Microcystis spp*, *Anabaena spp*, and *Planktothrix spp*, and to a

lesser extent *Dolichospermum spp., Geitlerinema spp., Leptolyngbya spp., Pseudanabaena spp., Synechococcus spp., Spirulina spp., Phormidium spp., Nostoc spp., Oscillatoria spp.,* and *Radiocystis spp*. [4]. *Microcystis aeruginosa* is the most common species of cyanobacteria found in freshwaters around the globe and has been associated with a number of human, livestock, and wildlife poisoning [5, 6].

M. aeruginosa commonly produces microcystin-LR (MC-LR) (**Figure 1**) which is the most toxic and most prevalent of the over 100 identified variants [4, 5]. All MCs share a common structure including a cyclic heptapeptide containing three D-amino acids (alanine, glutamic acid, and methylaspartic acid), two "unusual" amino acids (N-methyldehydroalanine and 3-amino-9- methoxy-2,6,8-trimethyl-10-phenyldeca-4,6-dienoic acid (ADDA)), and two variable L-amino acids (X and Z) [7]. MC-LR contains leucine and arginine in the X and Z positions, respectively, and accounts for 99% of total harmful algal blooms [8]. Other less common variants include MCLA, MC-YR, MC-RR, MC-LF, and MC-LW. They are believed to have lower toxicity than MC-LR [6]. MC-LR's biochemistry and toxicity are attributed to the ADDA moiety and its stereochemistry [9, 10]. Mechanistically, MC-LR targets hepatocytes in the liver and enters the cells by active transport aided by anion-transporting polypeptides [11, 12]. Next, the MC-LR binds strongly and irreversibly to serine or threonine protein phosphatases coded as PP1 and PP2A, which subsequently result in enzyme inhibition [13]. Given their importance in cell function and cell regulation, inhibition of PP1 and PP2A can result into hyper-phosphorylation of proteins and cytoskeletal filaments, which can induce apoptosis. MC-LR ingestion may also result in DNA damage, cell proliferation, and possible tumor promotion [12]. Acute toxicity can result in liver inflammation, hemorrhaging, and extensive hepatic bleeding. Death may occur due to liver failure at high or prolonged exposure.

MC-LRs are water-soluble and stable and demonstrate slow natural degradation (half-life = 10 weeks) in polluted water. The molecule is complex and heat-resistant making it toxic even after boiling. Although hard to remove by conventional water

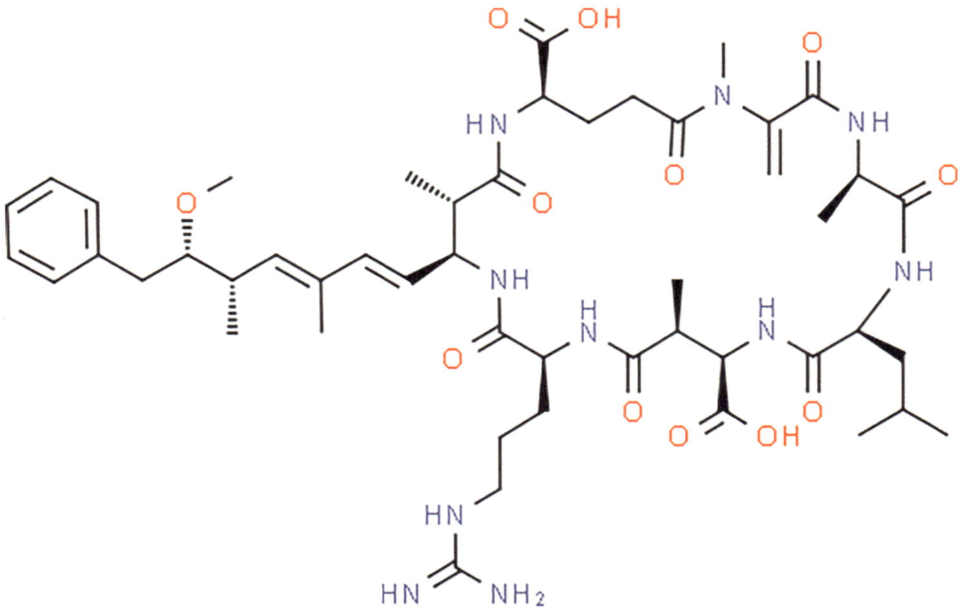

Figure 1.
General structure of microcystin-LR.

treatment, MC can rapidly degrade when exposed to UV radiation with wavelengths close to the absorption peak. Due to the presence of carboxyl, amino, and acylamino groups, MCs have been observed to ionize at temperatures above >40°C and in extreme acid-base media (pH <1.0 or pH >9.0) [9, 14, 15].

The distribution of MC in the US is a serious environmental health problem. Jenssen [16] has reported a wide range of MC concentration (12.5–225.6 μg/L) in multiple US communities. Environmental problems in the Wood County (West Virginia) and Mercer County (Ohio) closely reflect the national situation. Water quality data monitored between 2015 and 2019 by the EPA revealed that MC load in the Grand Lake St. Mary ranged from 0.0 to 79.7 μg/L, compared with the tolerable limit of 1.0 μg/L [17]. Similar data have been reported concerning the Ohio River Valley in West Virginia [16]. As mentioned before, UV exposure and electrocoagulation (EC) are useful methods for MC removal because of the capability to split their C–N bonds using electrical energy [15, 18, 19].

Recently, Folcik and Pillai [14] demonstrated the effectiveness of high-energy electron technology (advanced oxidation-reduction process) in degrading MC pollutants. The technology utilized accelerators to generate highly energetic electrons from regular electricity to create redox species to damage contaminants [20]. Similar examples of radiation technologies employed ^{60}Co gamma rays to inactivate MC multiplication [21–23]. Despite their effectiveness, these technologies are expensive and hi-tech and generally lack practical applications. Nevertheless, one of the techniques that are growing in popularity for MC decomposition is electrocoagulation [15].

Electrocoagulation (EC) employs the principles of electrochemistry for water treatment. It involves sacrificial corrosion of the electrodes (anode) to release active coagulant precursors (e.g., Al^{3+} or Fe^{2+}) into solution. At the cathode, hydrogen gas evolves from electrolytic reactions. EC equipment can theoretically be scaled for any size and is not too difficult to operate. Recent technical improvements combined with a growing need for small-scale water treatment facilities have amplified interest in EC applications. Nonetheless, only a few studies have focused on the question of scale to demonstrate how laboratory filtration can be upgraded to municipal treatments. In addition, elucidating the key components that control EC production and MC removal efficiency is of paramount interest. Some of the factors that require illumination include current density, electrical polarity, and acid-base equilibria [24]. We hypothesize that a coupled UV-electrocoagulation process will completely remove MC from contaminated drinking water. We also predict that laboratory EC techniques are scalable to municipal purification cognizant that strong water treatment oxidizers like ozone are obtainable from the system's vapor-ion plasma. The aim of this study is to (1) examine the operability and efficiency of cheap laboratory EC units for removing MC from drinking water, (2) test the scalability of laboratory EC filtration to municipal treatments, (3) evaluate the efficiency of the EC results against commercial water filtration (granulated activated carbon), and (4) examine the effects of radiation density, electrical polarity, pH, and temperature on the ionization of MC pollutants. The study will raise questions about electrocoagulation and industrial chemical recycling.

The chapter is structured in the following way. The first part reviews the literature on MC decomposition followed by description of the EC technique including the key components of the electrical units, electrodes activation, and reaction chemistry. The second section discusses the EC methodology followed by data generation, data analysis, and EC scalability. The final part examines the factors controlling EC physics,

including radiation density, electrical polarity, pH, and temperature. The final section also discusses the economy of the new EC method.

2. Materials and methods

2.1 Equipment and raw materials

The basic raw material is surface and groundwater samples from the Mid-Ohio River Valley in Parkersburg (West Virginia) and untreated water from the Grand St. Mary's Lake in Celina (Ohio). The illustration in **Figure 2a** represents the experimental design describing the general process of treating contaminated water using electrical energy. **Figure 2b** displays the laboratory-built UV-ozone water purification prototype, consisting of a 100-gallon plastic tank batch reactor fitted with an ionized nitrogen-oxygen (NI-OX) generator. The device is also fitted with a small fractional horsepower delivery compressor and 1-μm electron separation porous cellulose fiber water filter. Using basic engineering ideas, the unit was powered by a 110-V electrical source with the generator fastened to the tank cover and connected to a 1-μm polarized polymer filter suspended 10 inches above the inside base of the tank. The filter was connected to a fine bubble aeration diffuser using a half-inch poly tubing designed to eliminate debris, suspended solids, and microcystins pollutants. The principal component of the generator is a UV radiation lamp (λ = 155 nm) capable of

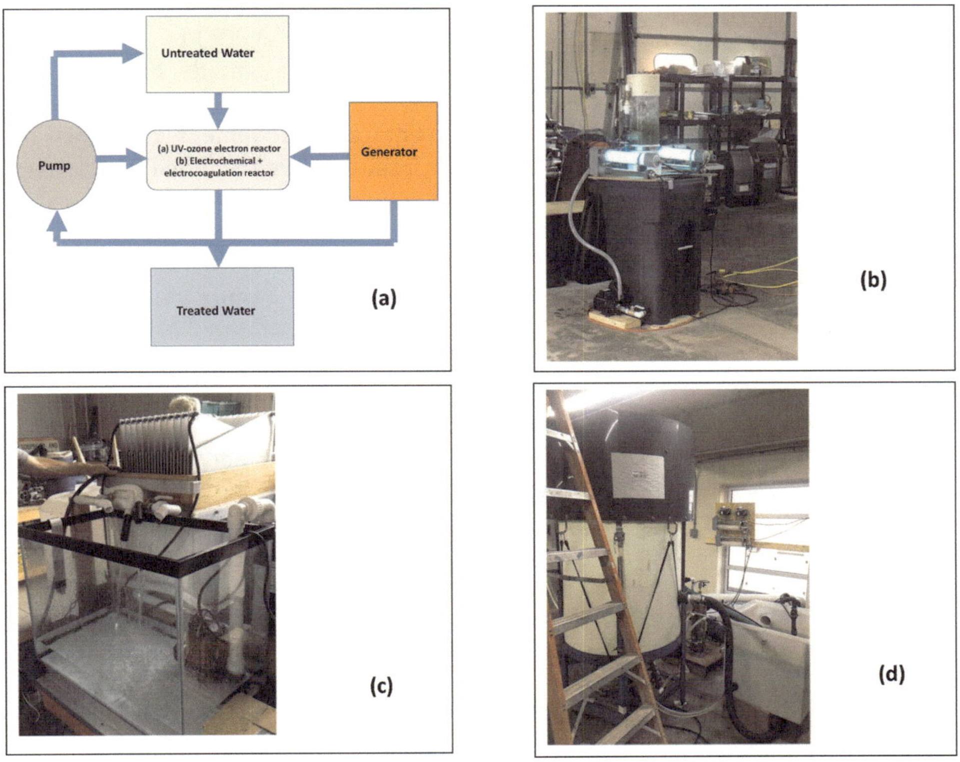

Figure 2.
(a) Schematic of the electrocoagulation purification system; (b) UV-ozone filtration unit; (c) Electrocoagulation unit; and (d) Coupled UV-ozone-electrocoagulation filtration system.

splitting ambient gases (e.g., O_2 and N_2) into monoatomic-charged particles using ultraviolet ionizing energy and magnetic emission.

Figure 2c is a modified version of the prototype in **Figure 2b** designed to suit field conditions. It consists of a 400-gallon steel tank powered by a high-amperage (250 A), low-voltage generator (40 V) constructed to provide energy via switching polarity from direct current electric discharge. A characteristic component is 34 pairs of submerged anode and cathode crosslinked aluminum electrodes secured over a steel tank (**Figure 2c**). Crosslinked electrodes are the main reason for the (switching) reverse polarity. The coagulator works by establishing an intense electromagnetic field creating simultaneous oxidation-reduction reactions. An attached high-pressure pump was designed to channel polluted water over the metal plate contact areas. Treated water was pumped into a clean glass tank before samples are drawn for testing.

The third equipment (**Figure 2d**) is a coupled UV-EC (UV-ozone-EC) system, designed to mimic an industrial treatment system, with capacity to purify approximately 10,000 gallons of contaminated water within 12 h. **Table 1** shows a summary of technical characteristics of the EC processing system.

2.2 Experimental methods

Raw water and treated water samples were tested at the Industrial Chemical Laboratories, LLC (ICL) of Denver (CO). ICL is a specialized facility for testing chemical and biological pollutants in drinking water and wastewater including cyanobacteria. The samples were tested every 10 min for 90 min and analyzed using the Agilent 1100 tandem high-performance liquid chromatography-mass spectrometer (HPLC-MS). The glassware was thoroughly washed and rinsed with methanol and distilled water to prevent cross-contamination. The samples were first filtered with Whatman filter paper (1.2 µm) and chilled overnight at −20°C to dilute concentration of the pollutants. The filtrate was dissolved in 400 µL methanol and treated with a 2 mg/L sodium thiosulfate and acidified with trifluoroacetic acid (TFA, 0.1%, v/v),

	Scale	**Raw Water Sample (Gallons)**	**Pump Horsepower**	**Energy from Generator**	**Minimum Time of Microcystin Decomposition (Minutes)**	**Maximum Time of Microcystin Decomposition (Minutes)**	
Water treatment method				**Voltage (Volts)**	**Electrical current (Amperes)**		
UV-ozone ionization	Laboratory	90	1.5	9	80	20	50
Electro-coagulation	Field	400	2.5	24	160	10	30
UV-ozone-electro-coagulation	Municipal	10,000	3.5	32	250	50	50
Activated carbon	Field	5	Not applicable	Not applicable	Not applicable	60	60

Table 1.
Characteristics of the water treatment techniques.

concentrated via solid phase cartridges (SuperClean LC-18, 3 mL tube), and eluted with 15 mL of 0.1% TFA in methanol. Aliquots of 20 μL were injected into system's column (150 × 4.60 mm) at a flow rate of 1 mL/min at 30°C column temperature. The mobile phase consisted of H_2O plus 0.05% TFA and acetonitrile plus 0.05% TFA with a linear increase from 30 to 70% of the latter between 0 and 30 min. Chromatograms were recorded at 238 nm based on the literature. UV spectra and all chromatographic peaks were examined and compared to spectra standards of MC moieties. Peaks possessing the UV spectrum characteristic for MCs were quantified using a calibration curve. Unidentified peaks possessing the UV spectrum characteristic for MC but not matching the retention time of the standards were determined as MC-LR equivalents with a detection limit of 0.01 μg/L [25]. To understand seasonal variations, MC distribution was measured seven consecutive days in spring, summer, and fall and a regional mean calculated (**Table 2**).

The UV-ozone ionization reaction process was produced following the reactions below. Charged nitrogen particles were activated to release of free electrons (e^-) to accelerate oxygen ionization:

$$N_2 \frac{UV}{MagE} N^{\alpha+} - N^{\alpha-} \frac{e^-}{O_2 ionized} (\text{free electron } [e^-] \text{ plus accelerating ionized } O_2). \tag{1}$$

The oxygen radiation produced ozone vapor, ionized ozone, and superoxide ions and dissociated into more singlet oxygen (Eq. 2), resulting in a chain reaction of high-energy ionized oxygen in (Eq. 3):

$$O_2 \frac{UV}{MagE} O_3 \frac{UV}{MagE} \rightarrow O_3^{+} \leftrightarrow O_3^{-} \frac{UV}{MagE} \rightarrow O_2^{-} \text{ (superoxide ion)} \tag{2}$$

$$O - O^- - +xO^- - O^-(\sin glet) - \frac{UV + electrons}{MagE} {}^-O - {}^-O - {}^-O(-O_x^-) - {}^-O - (chain) \tag{3}$$

The reaction of singlet oxygen (or chained ionized oxygen) with water was generated to produce high concentrations of hydrogen peroxide and/or hydroxide ions as saturated water produces excess peroxyl-reactive (oxidizing, disinfecting, and coagulating) ionized water in the subsequent reactions:

$$xO_2^- + H_2O \frac{e^-}{stream} \rightarrow xH_2O_2 \leftrightarrow xOH^- + xOH_2^- (scavenger) \tag{4}$$

$$O_2^- + H_2O \frac{e^-}{MagE_{\Delta T}} \rightarrow H_2O_2 \tag{5}$$

Thermal reaction of hydrogen peroxide and ozone was created to release free electrons and trioxidanes, superoxide ions, and peroxone Eqs. (6)–(8). The charged nitrogen and superoxide ions in aqueous solution were designed to produce additional free electrons, dinitrogen tetraoxide (nitroxyl ions), and hydroxide ions toxic to cyanobacteria:

$$H_2O_2 + O_3 \frac{e^-}{MagE_{\Delta T}} \rightarrow H_2O_3 + O_2^- \tag{6}$$

DOI: http://dx.doi.org/10.5772/intechopen.105751

	Range of water sampling depth (m)	Temperature (oC)	pH (-)	Turbi-dity (NTU)	Total dissolved Solids (mg/L)	Total organic carbon (mg/L)	Total nitrogen (mg/L)	Total phosphorus (μg/L)	Microcystins turbidity (μg/L)
Ohio River (Parkers-burg, WV)	0.0–1.5	23.3	8.58	0.41	121.4	3.8	0.62	27.23	112.6
Grand Lake St. Mary's (Celina, OH)	0.0–1.5	22.9	8.55	0.67	126.7	3.1	6.88	301.91	147.5

Table 2.
Relationship between the microcystin and nutrient load.

$$O_3 + H_2O_2 \frac{e^-}{\Delta T} \rightarrow H_2O_5 \tag{7}$$

$$N_2^+ + 3O_2^- + 2H_2O \frac{e^-}{MagE_{\Delta T}} {}^-O_2 - N_2^+ - O_2^- + 4OH^- \tag{8}$$

The electrocoagulation reverse polarity reaction follows an electrolytic procedure [26]. The primary reactions at the anode and cathode are described in Eqs. (9)–(13):

$$2\mathrm{H_2O}(l) \rightarrow \mathrm{O_2}(g) + 4\mathrm{H^+}(aq) + 4\mathrm{e^-} \text{ Anode} \tag{9}$$

$$4\mathrm{H_2O}(l) + 4\mathrm{e^-} \rightarrow 2\mathrm{H_2}(g) + 4\mathrm{OH^-}(aq) \text{ Cathode} \tag{10}$$

$$6\mathrm{H_2O}(l) \rightarrow 2\mathrm{H_2}(g) + \mathrm{O_2}(g) + 4\mathrm{H^+}(aq) + 4\mathrm{OH^-}(aq) \qquad \text{Overall} \tag{11}$$

$$4\mathrm{H^+} + 4\mathrm{OH^-}(aq) \rightarrow 4\mathrm{H_2O}(l) \tag{12}$$

$$2\mathrm{H_2O}(l) \rightarrow 2\mathrm{H_2}(g) + \mathrm{O_2}(g) \tag{13}$$

While reductants (free electrons) are released from the anode, oxidants and flocculation aggregates (e.g., H_2O_2, $Al(OH)_3$, Al_2O_3) are generated at the cathode as shown in Eqs. (14)–(18):

$$Xe^- + H_2O \xrightarrow{\text{Metal Electrodes}} H_2\uparrow + O_2\uparrow \xrightarrow{\text{Electrons}} O^- - O^- - O_x^-(Oxidation) \tag{14}$$

$$\xrightarrow{\text{Electrons}} OH^- + H_2O_2 \tag{15}$$

$$\xrightarrow{\text{Electrons}} Al(OH)_3\downarrow \tag{16}$$

$$\xrightarrow{\text{Electrons}} Al_2O_3\downarrow \tag{17}$$

$$\xrightarrow{\text{Electrons}} Electrophilic + Nucleophilic\ cleavage\ of\ C - F\ bonds \tag{18}$$

The pollutant removal efficiency (% r) was calculated using Eq. (19) as follows:

$$\%r = \left(\frac{C0 - Ct}{C0}\right) x\ 100\% \tag{19}$$

where C_O is the initial concentration of pollutant and C_t is the concentration of pollutant at time t.

To initialize the EC process and augment flocculent formation, about 80 g of potassium aluminum sulfate dodecahydrate ($KAl_2(SO_4)_2.12H_2O$) (potash alum) solution was used both as an electrolyte and coagulant following previous experiments by Johnson [26]. Finally, treated water samples were compared with data from a commercial gravity block ionic adsorption unit fitted with granular activated carbon filters and coated by silver-impregnated ceramic outer shells. The experimental results are discussed in the subsequent section.

3. Results and discussion

3.1 Electrocoagulation and scalability

The results in **Figure 3** displays microcystin (MC) response to laboratory UV exposure compared with field and municipal electrocoagulation (EC). To determine

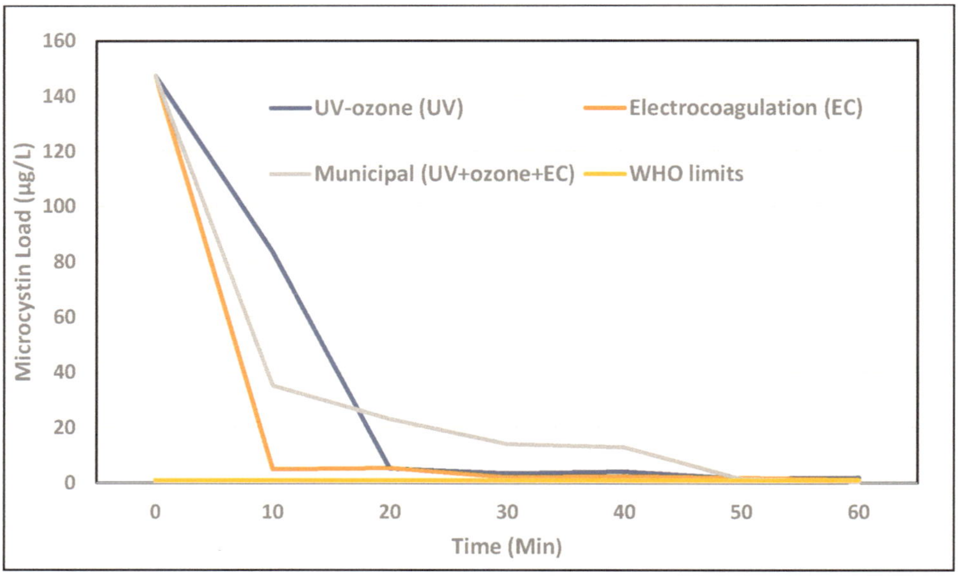

Figure 3.
Comparison of microcystin removal techniques.

the EC production efficiency, MC filtration data were compared with the WHO's maximum contaminant level of 1.0 µg/L. While the laboratory and field experiments decomposed MC within 10–20 min, the municipal system disrupted MC bonds after 50 min. Multiple reasons can explain this. The first one is a technical challenge. As expected, building and testing the 10,000-gallon reactor (**Table 1**) was more arduous than the 90-gallon reactor. The installation of high-intensity UV lamps to generate optimal radiation density in the larger reactor was particularly challenging. Another difficulty was how to generate maximum turbulence to aerate and circulate radiation. Although this was improved using bubble diffusers, predicting diffuser size was still time consuming, requiring several iterations. The reaction delays were also attributable to differences in surface energy interactions between radiation pulse and pollutant substrates. A recent study by Cavitt et al. [27] has shown that molecular bond disruption in aqueous media is controlled by many thermodynamic factors such as reaction rates, solvent volume, acid-base equilibria, and interfacial alignment of reactants versus products. Given similar temperature doses, reaction rates were better favored in the laboratory (90-gallon) reactor than its municipal (10,000-gallon) counterpart. The general conclusion is that MC bond disruption is easier in smaller reaction tanks than larger ones.

The aforementioned result notwithstanding the results in **Figure 3** shows a slight deviation between UV irradiation (20 min) and EC (electrolytic) treatments (10 min). Notice that UV irradiation is closely related to physical factors such as UV lamp size, pulse intensity, and radiation diffusion (15, 18), while EC is controlled by direct electrical vibration against C–N bonds. The aforementioned, therefore, is a reasonable explanation for the observed discrepancy. Another reason for the reaction delay is ozone deficiency possibly resulting from coupling glitches between the system's ionized nitrogen-oxygen (NI–OX) generator and its compressor (**Figure 2b**). This matter will be further investigated in subsequent studies. It is worth noting, however, that all the prototypes (**Figure 2**) decomposed MC molecules reasonably well. Notice for

instance, that the municipal EC unit destroyed MC by approximately 80% within the first 10 min, while the coupled UV-ozone treatment was even better at 95%. Studies such as Wolfe et al. [28], Langlais et al. [29], and Folcik et al. [14] have reported polar bond disruption from electron bombardment, free radical attack, and ozone and peroxone toxicity. Peroxone is a mixture between ozone and hydrogen peroxide Eqs. (1)-(8). The theoretical basis is that heavy oxidizing agents (e.g., peroxides, trioxidanes, and peroxones) can break down functional C–N bonds of microcystins [30, 31]. Previously, He et al. [32] reported the destruction of cyanobacteria using hydroxyl-free radicals. In addition, studies such as Yoo et al. [33] and Lui et al. [34] have demonstrated how low doses of peroxyl and nitroxyl ions can disrupt chemical bonds in molecular compounds. Results from this study quite closely reflect conclusions by previous researchers.

To validate MC decomposition data in **Figure 3**, UV treatments were matched against granular activated-carbon filtration data. The carbon filtration was from a commercial source and readily available. The results are displayed in **Figure 4a**. Two important things are observable from the outcomes. While the UV filtration disrupted C–N bonds within 20 min, the activated carbon produced treatments after 30 min. Still, the UV filtration outperformed the commercial granular carbon filtration by nearly 40%. This was expected knowing that UV radiation is more energetic in splitting C–N bonds. To predict the rate of MC removal, the data were subjected to a crude regression analysis. The curve followed a polynomial decay in the form of $y = 8.394x^2–88.635x + 221.97$; $R^2 = 0.9399$ (**Figure 4b**) confirming the robustness of UV radiation in removing MC pollutants within 1 h. This knowledge is relevant and applicable to wastewater treatment and chemical recycling.

3.2 Principal component analysis

This section discusses electrocoagulation (EC) principal components that control MC decomposition. The parameters below were considered important in the published literature [15]: (a) voltage (as proxy data for radiation density), (b) pH (acid-base equilibria), (c) electrical polarity (reverse polarity), and (d) temperature. The parametric data were derived using Eq. (19). The results are displayed in **Figure 5**. The municipal EC results were evaluated using the published data by Miao et al. [31] and further verified [35]. Two important points are observable here. First, MC removal increased with increasing radiation density. Second, there was a difference in optimal voltage density coincidental with maximum MC decomposition. While the field reactor completely removed MC (100% decomposition) at 24 V, the municipal reactor performed maximally at 32–40 V showing 95% pollutant removal (**Figure 5a**). The difference in energy dosage was attributed to the sheer size of the municipal reactor, which in turn resulted from generator adjustments to solve solvent turbulence and flocculent formation deficiencies.

As expected, no large differences were observed between the field and municipal reactions in terms of acid-base equilibria and thermodynamics (**Figure 5b** and **d**). The most significant conclusions are that (1) pH and temperature elevations are more favorable to MC decomposition; and (2) optimal pH for pollutant removal lies in the alkaline range with pH > 8.00. While the pH data strongly agreed with recent findings by Folcik and Pillai [14], it was strikingly contrary to previous conclusions by Bao et al. [16] whose studies on C–F decomposition was more productive in acidic media. It is possible to explain our findings in two ways. First, the coagulation "seeding agent" (i.e., potash alum ($KAl_2(SO_4)_2.12H_2O$) may have contributed to pH elevation.

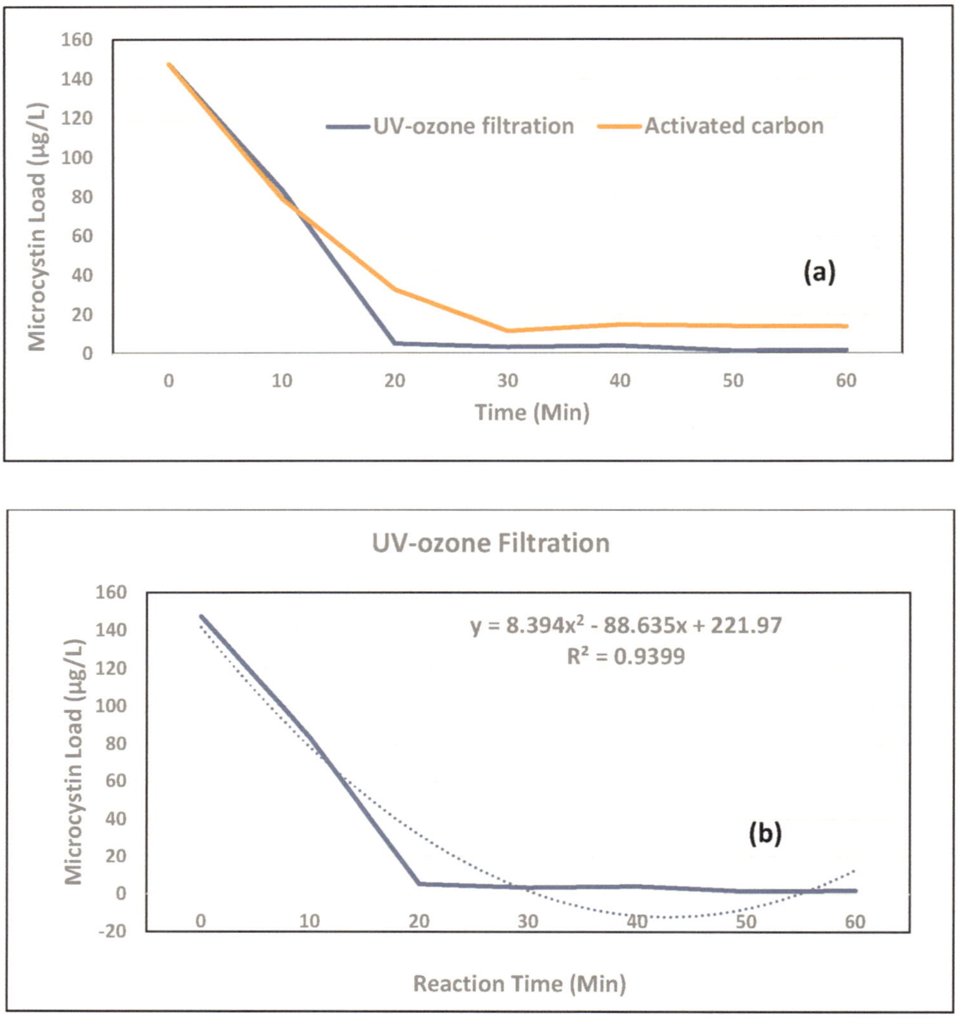

Figure 4.
(a) Comparison between UV radiation and granular activated-carbon removal of microcystins; and (b) Regression curve of the UV-ozone water treatment.

Secondly, the production of metal hydroxides such as $Al(OH)_3$ and $Fe(OH)_3$ from sacrificial anodes (aluminum and iron metals electrodes in the electrolytic cell (**Figure 2**)) may have produced more alkaline conditions. Thermodynamic effects on MC decomposition are well researched including published articles by Folcik and Pillai [14], Folcik et al. [36], and Wang et al. [37]. The conclusion here is that MC bonds are significantly disrupted at temperatures beyond 40°C. The results in **Figure 5d** quite closely matched some of the aforementioned findings. In this study, however, 30°C was observed as the starting point of MC decomposition with maximum disruptions encountered between 70 and 90°C. The difference may largely be due to high generator amperage (250 A) employed (Section 2.1).

Another important EC factor is electrical polarity. Previous studies such as Triantis et al. [38] and Gajda et al. [39] have reported limitations of conventional single-anode polarity in EC production. For this reason, we experimented a more robust switching polarity procedure using crosslinked aluminum electrodes with its energy from direct

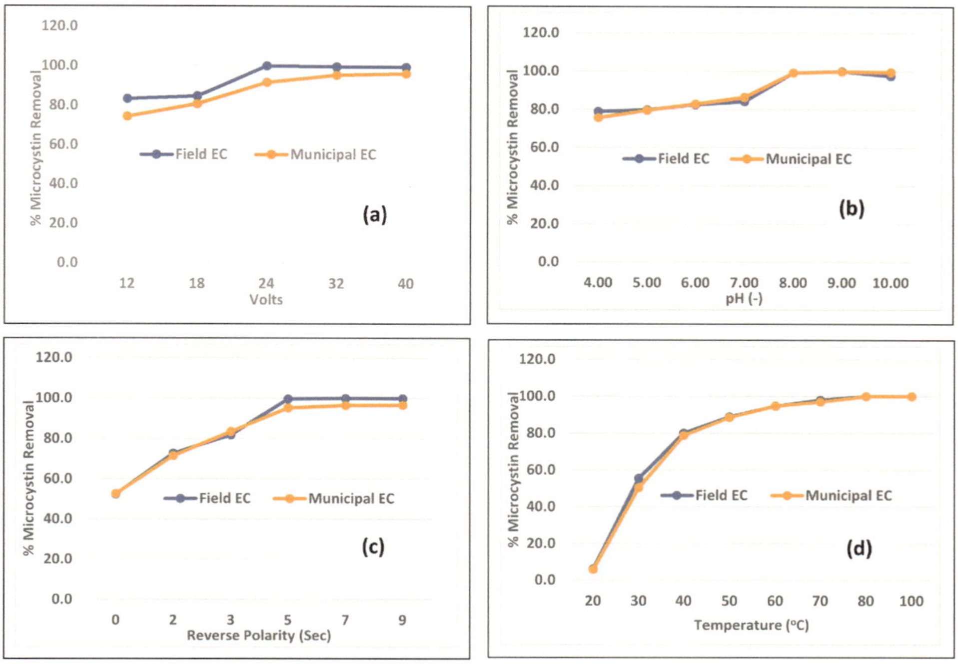

Figure 5.
Factors controlling electrocoagulation efficiency in decomposing microcystin (a) radiation density; (b) bond polarity; (c) pH; and (d) temperature.

current electrical discharge. Using a trial-and-error optimization approach, the reactor was "trained" to switch electrical current bombardments between the anode and cathode electrodes. The data in **Figure 5c** show that MC removal was maximum (100% removal) at every 5 s. The question is: Why is the switching polarity so important? The answer relates to two important things. First, optimizing the system saved time, power, and ultimately, cost. Second, the switching polarity ensured that the C–N bonds were attacked at both the anode and the cathode, contrary to conventional one-way electrical bond splitting. The dual attack against C–N bonds is a major reason for effective EC production. Notice, however, that the response from the large-size municipal reactor was inferior compared with its field counterpart. The discrepancy is still not easy to explain. However, operational problems such as electrode size determination for maximum flocculent distribution may be responsible for deviations. This is another topic worth investigating in future studies.

How does the new EC equipment compare with conventional community water treatment in terms of cost economics? To answer this question, the municipal prototype was "starved" of ozone and UV radiation, while extending treatments beyond 90 min. The goal was to examine whether the EC system would provide cheaper filtration compared with conventional treatments in the study area. The results are displayed in **Table 3**. The data show that while groundwater MC treatment was unimportant in West Virginia, the importance of surface water treatment was without question. The heavy pollution associated with both the Ohio River and Grand St. Mary's Lake (**Table 2**) is noticed. The data in **Table 3** show that the EC procedure was much cheaper than conventional membrane filtration or chemical disinfection. Specific to MC removal, the EC method was predicted to be nearly 800 times cheaper than conventional treatments at the Celina plant. On the basis of this study,

	Celina (Ohio) Grand Lake St. Mary's	Parkersburg (West Virginia)
Source of Drinking Water	Lake water	Groundwater (Ohio River Valley)
Scale of Water Treatment	Municipal	Municipal
Rate of Water Treatment	1000 gallons/minute	265 gallons/ minute
Cost of Water Treatment	$3.66/1000 gallons/ minute	$1.84/1000 gallons/minute
Cost of Microcystin Removal using Conventional Techniques (Aeration, bio-digestion & membrane filtration)	$0.37/1000 gallons/ minute	$0.00/1000 gallons/minute*
Cost of Microcystin Removal by Electrocoagulation	$0.04/1000 gallons/ minute	$0.04/1000 gallons/minute

**Parkersburg groundwater treatment is achieved by activated carbon filtration.*

Table 3.
Costs comparison between conventional water treatment and new electrocoagulation.

incorporating EC methods at conventional treatment plants has potential to both improve water treatment chemistry and save cost.

3.3 Further studies

This study has confirmed published reports that advanced EC methods are effective in removing MC pollutants from drinking water. However, some key topics remain to be investigated including (a) chemical elucidation of decomposed MC fragments, (b) scalability of UV lamps, ionized nitrogen-oxygen (NI-OX) generators and EC electrodes, (c) technical operability and cost assessments, and (d) applicability of the technique for wastewater treatment and chemical recycling. Some of these topics will be addressed in subsequent studies.

4. Conclusion

The removal of microscopic pollutants in drinking by electrocoagulation (EC) is becoming increasing popular because of the use of radiation energy in decomposing molecules that contain polar bonds including C–F and C–N bonds. The purpose of this study was to discuss three EC techniques for removing microcystins (MC) in contaminated drinking water at the Celina (OH) and Parkersburg (WV) treatment plants and to compare their effectiveness at the laboratory, field, and municipal scales. While the laboratory and field experiments employed UV-ozone and electrolytic cell filtration techniques, respectively, the municipal experiment applied a coupled UV-ozone and EC technique. To validate the effectiveness of the methods, the EC results were evaluated against a commercially available granular activated carbon filtration unit. The EC technique outperformed the activated carbon filtration by more than 40%. When the laboratory treatments were upscaled and tested at a municipal level, effectiveness of the technique declined by nearly 10–20% because of pulse dissipation from UV lamps, vapor-ion plasma underactivity, and limitation of membrane filters. We

confirmed previously published studies that pollutant coagulation and MC decomposition were affected by physical factors such as radiation density, reverse electrical polarity, pH, and temperature. These results have other applications in industrial wastewater treatment and chemical recycling.

Acknowledgements

The authors are very grateful to the Vienna City and Mercer County Councils for funding the chemical data analyses. We are also grateful to the biochemistry majors of Ohio Valley University for assisting coagulator test runs and data collection. Finally, we appreciate Dr. Matt Vergne of Lipscomb University, Nashville, Tennessee, for his critical review of the original manuscript.

Conflict of interest

The authors declare no competing financial interests.

Author details

Stephen Opoku-Duah[1]*, Dennis Johnson[2], Dan Blair[3] and Jeff Dimick[3]

1 Department of Chemistry and Biochemistry, Lipscomb University, One University Park Drive, Nashville, USA

2 Katharos Scientific LLC, Colorado Springs, USA

3 Oklahoma Christian University, Oklahoma City, USA

*Address all correspondence to: sopokuduah@lipscomb.edu

References

[1] Yu S, Zhao N, Zi X. The relationship between cyanotoxin (microcystin, MC) in pond-ditch water and primary liver cancer in China. Chinese Journal of Oncology. 2001;**23**(2):96-99

[2] Wiegand C, Pflugmacher S. Ecotoxicological effects of selected cyanobacterial secondary metabolites a short review. Toxicology and Applied Pharmacology. 2005;**203**(3):201-218. DOI: 10.1016/J.TAAP.2004.11.002

[3] Preece EP, Hardy FJ, Moore BC, Bryan M. A review of microcystin detections in Estuarine and Marine waters: Environmental implications and human health risk. Harmful Algae. 2017; **61**:31-45. DOI: 10.1016/j.hal.2016.11.006

[4] Buratti FM, Mangnelli M, Vichi M, Stafenelli M, Testai E, Funari E. Cyanotoxins: Producing organisms, occurrence, toxicity, mechanism of action and human health toxicological risk evaluation. Archives of Toxicology. 2017;**91**(3):1049-1130. DOI: 10.1007/s00204-016-1913-6

[5] Tanabe Y, Hodoki Y, Sano T, Tada K, Watanabe MM. Adaptation of the freshwater bloom-forming cyanobacterium Microcystis aeruginosa to brackish water is driven by recent horizontal transfer of sucrose genes. Frontiers in Microbiology. 2018;**9**:1150. DOI: 10.3389/fmicb.2018.01150

[6] Kubickova B, Babica P, Hilscherova K, Sindlerova L. Effects of cyanobacterial toxins on the human gastrointestinal tract and the mucosal inmate immune system. Environmental Sciences Europe. 2019;**31**(31):1-27

[7] Sharma VK, Triantis T, Antoniou MG, Xuexiang H. Destruction of microcystins by conventional and advanced oxidation processes: A review. Separation and Purification Technology. 2012;**91**:3-7. DOI: 10.1016/j.seppur.2012.02.018

[8] Vasconcelos V, Sivonen K, Evans WR, Carmichael W. Hepatotoxic microcystin diversity in cyanobacterial blooms collected in Portuguese freshwaters. Water Research. 1996; **30**(10):2377-2384. DOI: 10.1016/0043-1354(96)00152-2

[9] Tsuji K, Naito S, Kondo F, Ishikawa N, Watanabe MF, Suzuki M, et al. Stability of microcystins from cyanobacteria: Effect of light on decomposition and isomerization. 1994. Environmental Science and Technology. 1994;**28**(1): 173-177. DOI: 10.1021/es00050a024

[10] Song W, De La Cruz AA, Rein K, O'Shea KE. Ultrasonically induced degradation of microcystin-LR and RR: Identification of products, effect of pH, formation and destruction of peroxides. Environmental Science and Technology. 2006;**40**(12):3941-3946. DOI: 10.1021/es0521730

[11] Fischer WJ, Altheimera S, Cattori V, Meier PJ, Dietrich DR, Hagenbuch B. Organic anion transporting polypeptides expressed in liver and brain mediate uptake of microcystin. Toxicology and Applied Pharmacology. 2005;**203**(3): 257-263. DOI: 10.1016/J.TAAP.2004.08.012

[12] Valério E, Chaves S, Tenreiro R. Diversity and impact of prokaryotic toxins on aquatic environments: A review toxins. Molecular Diversity Preservation International. 2010;**2**(12): 2359-2410. DOI: 10.3390/toxins2102359

[13] Craig M, Luu HA, McCready TL, Holmes CFB, Williams D, Andersen RJ.

Molecular mechanisms underlying the interaction of motuporin and microcystins with type-1 and type-2A protein phosphatases. Biochemistry and Cell Biology. 1996;**74**:569-578. DOI: 10.1139/o96-061

[14] Folcik AM, Pillai SD. A critical review of ionizing irradiation technologies for the remediation of waters containing microcystin-LR and M. Physics and Chemistry. 2020; **2020**(177):109128

[15] Hakizimana JN, Gourich B, Chafi M, Stiriba Y, Vial C, Drogui P, et al. Electrocoagulation process in water treatment: A review of electrocoagulation modeling approaches. Desalination. 2017;**404**:1-21

[16] Jenssen EM-L. Cyanobacterial peptides beyond microcystins–a review on co-occurrence, toxicity, and challenges for risk assessment. Water Research. 2019;**151**:488-499

[17] EPA-Ohio. Ohio algal toxins results from Lake Erie, State Park Beaches, Inland Lakes and public water supplies. (Algal Toxins Microsoft Excel Spreadsheet data records). 2019;**9**:34-78. Cincinnati, OH

[18] Bao J, Yu W-J, Liu Y, Wang X, Lui Z-Q, Duan Y-F. Removal of perfluoroalkanesulfonic acids (PFSAs) from synthetic and natural groundwater by electrocoagulation. Chemosphere. 2020;**248**:125951

[19] Rutberg PG, Kolikov VA, Kurochkin VE, Panina LK, Rutberg AP. Electrical discharges and the prolonged microbial resistance of water. IEEE Transactions on Plasma Science. 2007; **35**(4):1111-1118

[20] Miller RB. Electronic Irradiation of Foods: An Introduction To Food Irradiation. 1ed ed. Boston, MA: Springer; 2005. DOI: 10.1007/0-387-28386-2_1

[21] Bhatia SS, Pillai SD. A comparative analysis of the metabolomic response of electron beam inactivated E. coli O26: H11 and Salmonella typhimurium ATCC 13311. Frontiers in Microbiology. 2019; **10**:694. DOI: 10.3389/fmicb.2019.00694

[22] Jesudhasan PR, McReynolds JL, Bryd AJ, He H, Genovese KJ, Droleskey R, et al. Electron-beam inactivated vaccine against Salmonella enteritidis colonization in molting hens. Avian Diseases. 2015;**59**:165-170. DOI: 10.1637/10917-081014-ResNoteR

[23] Rocha JN, Dangott LJ, Mwangi W, Alaniz RC, Bordin AI, Cywes-Bentley C, et al. PNAG-specific equine IgG 1 mediates significantly greater opsonization and killing of Prescottella equi (formerly Rhodococcus equi) than does IgG 4/7. Vaccine. 2019; **37**:1142-1150. DOI: 10.1016/j.vaccine.2019.01.028

[24] Londono-Zuluanga C, Jameel H, Gonzalez RW, Lucia L. Crustacean shell-based biosorption water remediation platforms: Status and Perspectives. Journal of Environmental Management. 2019;**231**:757-762

[25] Addico GND, Hardege JD, Kohoutek J, deGraft-Johnson KAA, Babica P. Cyanobacteria and microcystin contamination in untreated and treated drinking water in Ghana. Advances in Oceanography and Limnology. 2017;**8**: 92-106

[26] Johnson D. LectroCell reactor system for drinking water purification: Design, chemical principles and operating manual. Techn. Reports—KATHAROS Scientific Inc. Reports, Dayspring Publ. CO, 2e. 2015:1-35

[27] Cavitt BT, Carlisle JG, Brooks RA, Scott LG, Patel P. Quantifying interfacial substrate interactions via surface energy analyses. STAR Protocols. 2021;**2**:100476

[28] Wolfe RL, Stewart MH, Scott KN, McGuire MJ. Inactivation of *Giardia muris* and indicator organisms seeded in surface water supplies by peroxone and ozone. Environmental Science and Technology. 1989;**23**:744-745

[29] Bablon G, Bellamy WD, Bourbigot M-M, Daniel FB, Erb F, et al. Practical applications of ozone: Principles and case studies. In: Langlais B, Reckhow DA, Brink DR, editors. Ozone in Water Treatment – Applications and Engineering. Lynden, USA: AWWA Lewis Publishers, Inc.; 1991. pp. 184-222

[30] Kim S-H, Moon B-Y, Lee H-I. Effects of pH and dosage on pollutant removal and floc structure during coagulation. Microchemical Journal. 2001;**68**(2–3): 197-203

[31] Miao H-F, Qin F, Tao G-J, Tao W-Y, Ruan W-Q. Detoxification and degradation of microcystin-LR and –RR by ozonation. Chemosphere. 2010;**79**: 355-361

[32] He X, Lui Y-L, Conklin A, Westrick J, Weavers LK, Dionysius DD, et al. Toxic cyanobacteria and drinking water: Impacts, detection, and treatment. Harmful Algae. 2016;**54**: 174-193

[33] Yoo S, Carmichael W, Hoehn R, Hrudey S. Cyanobacterial (Blue-Green Algal) Toxins: A Resource Guide. Denver CO: American Water Works Association Research Foundation; 1995. pp. 218-229

[34] Lui H, Zhao X, Qu J. Electrocoagulation in water treatment. In: Comninellis C, Chen G, editors. Electrochemistry for the Environment. New York, USA: Springer Int. Publishing, AG.; 2009. pp. 245-262. DOI: 10:1007/978-0-387-68318-8_10

[35] Opoku-Duah S, Johnson D. Removal of perfluorooctanoic acid and microcystins from drinking water by electrocoagulation. Journal of Chemistry. 2020;**10**:1-10. DOI: 10.1155/2020/1836264

[36] Folcik AM, Klemashevich C, Pillai SD. Response of Microcystis aeruginosa and Microcystin-LR to electron beam irradiation doses. Radiation Physics and Chemistry. 2021; **186**. DOI: 10.1016/j.radphyschem.2021.109534

[37] Wang H, Qu G, Gan Y, Zhang Z, Li R, Wang T. Elimination of microcystis aeruginosa in water via dielectric barrier discharge plasma: Efficacy, mechanism and toxin release. Journal of Hazardous Materials. 2022;**422**:126956. DOI: 10.1016/j.jhazmat.2021.126956

[38] Triantis TM, Fotiou T, Kaloudis T. Photocatalytic degradation and mineralization of microcystin-LR under UV-A, solar and visible light using nanostructured nitrogen doped TiO2. Journal of Hazardous Materials. 2012; **211-212**:196-202

[39] Gajda I, Greenman J, Mehuish C, Ieropoulos I. Electricity and disinfectant production from wastewater: Microbial fuel-cells as a self-powered electrolyser. Scientific Reports. 2016;**6**:1-9

Chapter 3

Cyanobacteria as the Source of Antioxidants

Rashi Tyagi, Pankaj Kumar Singh and Archana Tiwari

Abstract

The present-day scenario in the health sector calls for alternative medicine sources with no risk of resistance, effective in the mode of action, and eco-friendly. Cyanobacteria are microbial factories for a wide range of products. They are reservoirs of bioactive compounds which have the potential to act as precursors of novel drug molecules. A plethora of algae have been documented for their therapeutic abilities in treating diseases. A plethora of antioxidative compounds along with enzymes are present in cyanobacteria, possessing applications in nutraceuticals and cosmeceuticals, which is quite evident from the products available in the market. This chapter highlights the significant leads in the area of cyanobacteria-based antioxidants. A sustainable approach to envisaging cyanobacteria as competent antioxidants can open new doors in prevention, treatment, and control of a plethora of diseases.

Keywords: algae, antioxidants, cyanobacteria, cosmeceuticals, nutraceuticals

1. Introduction

Cyanobacteria exist in various habitats that are exposed to various adverse environmental variables, such as ultraviolet light, salinity, climate, and food supplements. Algae are multicellular creatures found in freshwater, saltwater, and marine environments. They synthesize a wide range of metabolites to acclimate to these demanding environments quickly [1]. Cyanobacteria's antioxidants can be used in the pharmaceutical and medical fields. The search for safe antioxidants derived from natural sources is currently generating interest on a global scale. Algae could biogenically create, consume, collect, and develop a wide variety of metabolites [2]. The agricultural, medicinal, pharmaceutical, food, nutritional, cosmetic, and other industries employ algae. In the absence of light, they can also grow under heterotrophic conditions by utilizing an organic carbon substrate as an energy source [3].

The existence of several proterozoic oil deposits is related to cyanobacterial activity. Additionally, they are significant suppliers of nitrogen fertilizer for growing rice and beans. Throughout the planet's history, cyanobacteria also played a major role in determining ecological change and evolution. Many cyanobacteria produce the oxygen atmosphere on which we rely. Before it, the atmosphere's chemistry was significantly different and unsuitable for modern species.

The nascent, most diversified, and wide cluster of photosynthetic prokaryotes known as cyanobacteria (blue-green algae) exhibits similarities to green plant life

in oxygenic photosynthesis and to Gram-negative bacteria in the cellular organization [4]. Almost all terrestrial and aquatic freshwater and marine habitats support the growth and colonization of blue-green algae, which adapt to diverse ecological circumstances [5]. Microalgae exist as a standard source of bioactive chemicals and have been used in various pharmacological applications due to their richness in primary and secondary metabolites [6]. Bioactive substances are physiologically active molecules that can either benefit or harm a living thing, tissue, or cell when present in small amounts [7].

Proteins called antioxidant enzymes to play a catalytic role in converting reactive oxygen species (ROS) and their byproducts into stable, harmless compounds, making them the most effective protection against oxidative stress-related cell damage. Antioxidant enzymes can stabilize or inactivate free radicals before damaging cellular components. They work by lowering the free radicals' energy or by sacrificing part of their electrons for its use, making the radicals more stable. To lessen the harm by free radicals, they may also halt the oxidative chain reaction. Over the last 10 years, numerous studies have been conducted on the advantages of antioxidant enzymes. Antioxidants protect cells against the harm caused by free ions, even when their concentration is lower than the substance being oxidized [8]. However, various harmful side effects, including cancer and liver damage, are associated with these synthetic antioxidants. As a result, scientists are looking for natural antioxidants that may be used in place of synthetic antioxidants in the food and pharmaceutical industries that are safe and efficient [9]. Our body's capacity to lower the risk of free radical-related health issues is made more tangible by minimizing exposure to free radicals and increasing the intake of foods or supplements rich in antioxidant enzymes [8]. Therefore, antioxidant enzymes are vitally essential for preserving the best possible cellular and systemic health and well-being. Free radicals are included in the highly reactive oxygen-containing molecules known as reactive oxygen species (ROS). The hydroxyl radical, superoxide anion radical, hydrogen peroxide, singlet oxygen, nitric oxide, and chlorine ions radicals, and other lipid peroxides are examples of ROS. All have the potential to interact with cellular membranes, phospholipids, nucleic acids, proteins, enzymes, and other tiny molecules to cause cellular harm [8].

Algae's numerous bioactive chemicals are being examined. PUFA, sterols, terpenoids, carotenoids, and alkaloids are just a few functional chemical components found in the diverse group of organisms known as algae. These have been shown to protect against various diseases, including cancer [9].

2. Algae antioxidants

Algae are photoautotroph organisms. There is no damage to the structure, and it can produce the substances they need to defend itself against oxidation. They are a rich source of powerful antioxidants that can shield our bodies from the harmful effects of oxygen species created during regular bodily metabolism. Carotenoids and vitamin E (gamma-tocopherol) are two types of powerful fat-soluble antioxidants found in algae, whereas vitamins, phycobiliproteins (PBPs), and polyphenols are adequate water-soluble antioxidants [9].

As naturally obtained bioactive compounds with a wide variety of biological potencies, such as antibacterial, antiviral, antioxidant, and anti-inflammatory, cyanobacteria have attracted much attention [1]. Antioxidants and phycobiliproteins (PBPs), which are cyanobacteria's distinctive photosynthetic pigments, are thought

to be abundant in cyanobacteria. In particular, these pigments have been exploited as organic coloring replacements in nutritive, cosmetic, and pharmaceutical products. Due to their fluorescent qualities, PBPs are also utilized in the branch of immunology [2]. Phycobiliproteins are highly effective fluorescent substances due to their distinctive characteristics of high molar absorbance coefficients, high fluorescence quantum yield, big stokes shift, high oligomer stability, and high photostability [10]. The primary endogenous damage to the biological system is caused by free radicals generated during oxidative stress. This kind of damage is frequently linked to several degenerative diseases and disorders, including cancer, cardiovascular disease, aging, and immune function loss. Free radicals are the main factor in lipid oxidation, the process by which food degrades and finally loses its properties to sense it through sensory organs and edibility, in addition to harming live cells [11]. Many individuals use antioxidants in the form of commercial food additives, which are produced synthetically and may contain significant amounts of preservatives, to combat the effects of oxidative stress [12]. However, most antioxidant sources identified to date compete with conventional meals and commodities.

Most biologically active compounds in algae, including pigments like carotene, astaxanthin, lutein, zeaxanthin, and phycobiliproteins [13], exhibit both anti-inflammatory and antioxidant properties [14]. One of the key factors driving the hunt for bioactive substances like anti-inflammatory kinetic molecules from natural sources such as microalgae is the rising demand for medications with few adverse effects. The cell that showed anti-inflammatory action will accumulate metabolites from the various microalgae. Several research has already shown the chemical makeup, structural details, and biosynthesis routes of the bioactive substances displaying anti-inflammatory chemicals produced by microalgae [15]. Proteins, phycobiliproteins, flavonoids, carotenoids like astaxanthin and lutein, and the fatty acids DHA, EPA, and SPs produced by metabolically active microalgae species are all present examples of substances with anti-inflammatory properties [2]. To be a valuable target product, these bioactive compounds must fulfill two requirements: (1) they must accumulate in relatively large amounts in cultures grown under standard test conditions throughout commercial production, and (2) they should continuously be overexpressed as an algal reaction to unpleasant development surroundings or when exposed to the synthetic or actual pressure. This can be achieved by differing the circumstances, like changing the physicochemical boundaries and the retention of supplements, as well as changes in temperature, pH, light quality, and irradiance [16]. The species of algae and the growing circumstances significantly impact the generation of anti-inflammatory chemicals [14]. Only a peptide from *P. tricornutum* with anti-inflammatory characteristics made it to market. Carotenoids, an algae stain, are discovered to positively influence immune response modulations and anti-inflammatory cellular response pathways [2]. *H. pluvialis* microalgae produce the carotenoid astaxanthin, which has strong anti-inflammatory properties [14]. One extremophilic microalga, *D. salina*, is used in industry to create a valuable substance with anti-inflammatory properties [17]. Microalgae-produced compounds have also been shown to have antioxidant as well as anti-inflammatory activity. Microalgal anti-inflammatory compounds known as sugars have also demonstrated antioxidant potential. Several excellent reviews have been published recently to discuss their uses and advantages for human health [2]. The antioxidant microalgae sugars extracted from *Porphyridium* [18] and *Rhodella* [19] are two excellent patterns.

Cyanobacteria inherently produce ROS throughout the photosynthetic activity. These species are created by abiotic causes such as ultraviolet radiation, osmotic disturbances, desiccation, and heat. Multiple strategies are needed for cyanobacteria

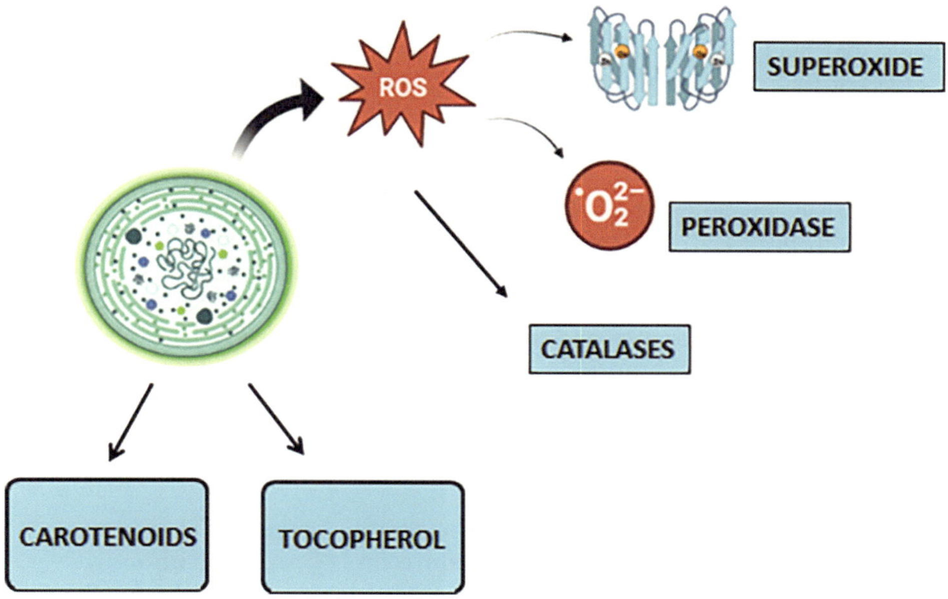

Figure 1.
ROS removing bioactive compounds.

to avoid the inhibitory effects of harsh conditions. By reducing the amount of energy lost during the photosynthetic process, they can reduce the generation of ROS. One approach uses the carotenoid zeaxanthin to non-photochemically quench (NPQ) excitation energy through photosystem II [20]. Cyanobacteria remove ROS using various bioactive compounds, as mentioned in **Figure 1**, and their genetic relationship can be elucidated using various methods of molecular phylogenetics [21]. Although peroxidases and catalases accelerate the removal of peroxides (such as H_2O_2 and R-O-O-H)13, SORs and SODs eliminate superoxide free radicals (O_2). These O_2 molecules are produced by photosynthetic and respiratory electron transport chains12, as well as extracellular processes on the cell surface. This promotes various processes like a ferrous deposition, cell signaling, and growth; however, if O_2 is allowed to accumulate inside the cell, it reacts with solvent-exposed 4Fe-4S clusters in proteins, including those needed for amino acid biosynthesis17 and photosynthesis18, resulting in Fenton reactants, which can eventually cause cell death. Therefore, SODs and SORs are discovered in all three branches – Eukarya, Archaea, and Bacteria [21].

Some of the important antioxidants found in algal species are listed below in **Table 1**.

S.no.	Algal species	Antioxidants	References
1.	*Chlorella zofingiensis, Chaetoceros gracilis*	Exopolysaccharides	[22, 23]
2.	*Chlorella ellipsoidea*	Carotenoids (mainly violaxanthin)	[24]
3.	*Chlorella vulgaris*	Carotenoids (mainly lutein)	[24]
4.	*Chaetoceros calcitrans, C. gracilis*	Fucoxanthin	[23, 25]

S.no.	Algal species	Antioxidants	References
5.	*Codium fragile*	Siphonaxanthin	[26]
6.	*Cyanophora Paradoxa*	β-Cryptoxanthin	[27]
7.	*Dunaliella salina*	β-carotene	[28, 29]
8.	*Dunaliella tertiolecta*	Violaxanthin	[29, 30]
9.	*Haematococcus pluvialis*	Astaxanthin	[31]
10.	*Nannochloropsis oculata*	Sterols	[32]
11.	*Nannochloropsis salina,*	PUFA	[33]
12.	*Skeletonema* sp.,	PUFA	[34]
13.	*Chaetoceros* sp.,	PUFA	[35]
14.	*Thalassiosira weissflogii*	PUFA	[36]
15.	*Phaeodactylum tricornutum*	Sulfated Polysaccharides	[29, 37]
16.	*Porphyridium purpureum*	Zeaxanthin	[38]
17.	*Porphyridium cruentum*	Sulfolipids	[39]
18.	*Spirulina platensis*	C-Phycocyanin	[40]
19.	*Tribonema* sp.	Sulfated polysaccharides	[29, 41]

Table 1.
Antioxidants from algae.

3. Antioxidative enzymes

3.1 Catalases

Initially, cyanobacteria were partitioned into two forms, those that have and those that need ascorbate peroxidase [42]. This depended on the perception that the principal bunch searches H_2O_2 with a peroxidase utilizing a photo-reductant as an electron benefactor. Cyanobacteria have three types of catalases (**Figure 2**) which differ significantly in terms of their structure and amino acid sequence. Bernroitner [43] examined the presence of these three catalases in 44 cyanobacterial genomes and executed a phylogenetic exploration of the enzymatic activities. The findings show that while monofunctional heme-containing catalase (KatE) is the most common type of catalase found in bacteria, archaea, and eukarya, it is extremely rare in cyanobacteria. Only one complete KatE gene was found in *Nostoc punctiforme* PCC73102, whereas pseudogenes (incomplete or fusion genes) were found in *Nostoc sp.* PCC7120, *Cyanothece sp.* ATCC51142, *and Synechococcus elongatus* [43].

KatG bifunctional catalase/peroxidase has both catalases and peroxidase activity. Unlike KatE, it was found in a variety of known cyanobacterial genomes. Cyanobacterial KatGs are known to form a well-segregated clade in the evolutionary representation, implying that KatG evolved in cyanobacterial evolution [43]. Mn catalase (MnCat) is a di-manganese catalase that does not contain heme. Except for *Gloeobacter violaceus* PCC7421, all species have MnCat. It is thought to be found only in diazotrophic cyanobacteria, except for *Gloeobacter violaceus* PCC7421 [43] (**Figure 2**).

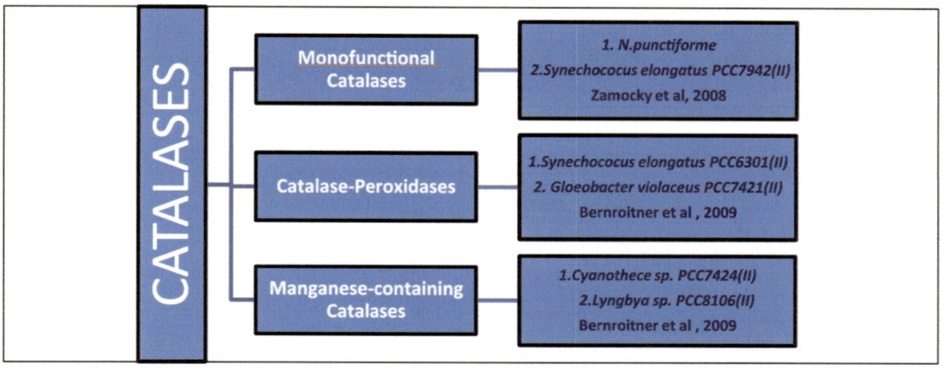

Figure 2.
Types of catalase enzyme.

3.2 Superoxide dismutase

SODs are common metalloenzymes and are classified into four types based on FeSOD, CuZnSOD, MnSOD, and NiSOD. All have metal redox-active centers that, respectively, include Fe (III), Cu (II), Zn (II), Mn (III), and Ni (II/III) at the active site [44]. Cyanobacteria have all four kinds of SOD, and many cyanobacterial species include more than one type of SOD [45]. It should be emphasized that some actinobacteria and archaea have a single gene that, depending on the environment, can either make FeSOD or MnSOD [46]. Cambialistic SOD refers to Fe/MnSOD that exhibits similar activity in Fe- and Mn-bound forms (Sheng et al. 2014). There are currently no known cabalistic Fe/MnSODs. While FeSOD and NiSOD or FeSOD and MnSOD are present in various other single-celled strains, the marine species of *Prochlorococcus* has only one NiSOD [44]. In contrast, strains that are heterocystous, heterotrichous, and flagellated exclusively have iron and manganese forms. Despite having comparable structural characteristics, FeSOD and MnSOD can be identified from one another by structural traits due to the existence of a transmembrane domain, residues mainly for some metals that differ between the two representations, and highly conserved residues found only in the manganese form [47]. Many investigations have found SODs to be involved in protective processes in cyanobacteria.

3.3 Peroxidases

Ascorbate peroxidase is essential for the detoxification of H_2O_2 in plants [48]. These enzymes convert H_2O_2 to monodehydroascorbate and water using ascorbate as the electron source. Ascorbate and dehydroascorbate are produced spontaneously by monodehydroascorbate. Dehydroascorbate reductase converts dehydroascorbate to ascorbate by using glutamine. NADPH-glutamine reductase then regenerates oxidized glutamine. This highlights the importance of the ascorbate-glutamine cycle in plant oxidative stress response. *Nostoc muscorum* PCC 7119 and *Synechococcus* PCC 6311 have both been found to contain ascorbate peroxidase-like activities, and dehydroascorbate reductase and glutamine reductase were both engaged in the regeneration of ascorbate and glutamine, respectively, in *Synechococcus* PCC 7942.

Peroxiredoxins (Prx-s), also known as alkyl-hydro peroxidases, are another widespread group of thiol-explicit cell reinforcement proteins that utilize thioredoxin and

other thiol-containing decreasing specialists as electron givers to diminish H_2O_2, alkyl hydroperoxides, and peroxynitrite [49]. It is believed that peroxiredoxins are crucial for decreasing endogenously produced ROS.

3.4 Superoxide reductases

All species interacting with air produce superoxide, or O2•-, which, depending on the biological environment, can function as a signaling agent, a poisonous lifeform, or a nontoxic precursor that breaks down spontaneously. Superoxide reductase (SOR) and superoxide dismutase (SOD) are two enzymes that limit their levels in vivo (SOD) [46]. SORs are simple enzymes with a sequence of 110–180 amino acids. SORs

S. No	Antioxidant compounds	Health benefits	References
1.	Astaxanthin	Influence the immune system and support cognitive health.	[50]
2.	β-Carotene	Preventative for breast cancer	[51]
3.	Bromophenol, Carrageenan	Inhibition of α glycosidase Antitumor, antiviral	[52]
4.	Carotenoids	Reduce the risk of cancer and eye disease.	[53]
5.	Chlorophyll	Prevent cancer and heal damaged skin	[54]
6.	Flavonoids	Anticancer activity prevents coronary heart disease	[55]
7.	Fucophlorethol	Chemopreventive	[56]
8.	Fucoidan	Improves hyperoxaluria Anticancer Protection against monogenic Disorder	[57, 58]
9.	Fucoxanthins	Ontogenesis	[59]
10.	Galactan sulfate	Antiviral	[60]
11.	Lutein	Anti-inflammatory	[51]
12.	Oligosaccharides	Anti-HIV	[61]
13.	Polyphenols	Vascular chemoprotection Antimicrobial α glycosidase inhibition	[55, 62]
14.	Phenolic functional groups and MAAs	Antiproliferative	[55, 63]
15.	Phycobiliproteins (PBPs)	Antiallergic, anti-inflammatory, neuroprotective.	[10, 64]
16.	Phlorotannin	reduce inflammation Kills bacteria	[65]
17.	Phycoerythrin	Reduce the effects of diabetes complications	[66]
18.	Xanthophylls	Neuroprotective	[50]

Table 2.
The advantages of algal antioxidants for health.

can be categorized in several ways. Neelaredoxins and desulfoferrodoxins have only one or two Fe atoms per polypeptide chain, respectively, which was the most distinguishing feature of these enzymes [44]. As a result, the most accurate classification for the procedure is to categorize them as 1Fe-SORs (neelaredoxins) and 2Fe-SORs (the desulfoferrodoxins). The most current exploration the methanoferrodoxin, a SOR from some methanogens with a domain harboring a $[4Fe\text{-}4S]^{2+/1+}$ cluster, may lead to an extension of this classification in the near future [46] (**Table 2**).

4. Non-enzymatic antioxidants

4.1 Carotenoids

Carotenoids are the most common and naturally occurring pigment. One such example is hydrophobic terpenoids. The polyene chain of carotenoids, which is made up of double bonds, gives them their coloration and the capacity to absorb photons of visible wavelengths. Both photosynthetic and non-photosynthetic species can produce carotenoids. Carotenes and xanthophylls are the two major categories of naturally occurring carotenoid algae [67]. Carotenes are linear or cyclic hydrocarbons, e.g., β-carotene and α-carotene. Oxygenated carotenoid derivatives are known as xanthophylls. The xanthophylls, violaxanthin, antheraxanthin, zeaxanthin, neoxanthin, and lutein produced by higher species are synthesized by green algae [67].

Freshwater pond cyanobacterial blooms emit a foul stench because of their adaptation to human-induced conditions exposed. These blooms of blue-green algae spread widely and produce cyanotoxin, poisonous to other creatures. However, these poisons have demonstrated potential properties as cancer treatments. Consider microcystins, numerous peptide toxins, including cryptophycins and anatoxin-A, have shown clinical effectiveness for various cancers [68].

Carotenoids, which are byproducts of photosynthesis and include carotene, xanthene, lutein, and lycopene, are often abundant in algae and cyanobacteria. As foragers of electron species with a singlet, or ROS, carotenoids and other terpenoids are crucial. Therefore, these scavengers are used as antioxidants to stop the growth of cancer cells. There are not many reports on carotenoids' ability to fight different types of cancer [69].

Astaxanthin and β-carotene, generated by *Haematococcus pluvialis* and *Dunaliella salina*, respectively, are two main carotenoids produced by microalgae. A vital component known as β-carotene is extensively looked for as a food coloring agent, for cosmetics addition, and as healthy food. It is frequently used in soft drinks, cheeses, butter, and margarine and is well-known for being safe and having health benefits due to its pro-vitamin A activity [70]. Astaxanthin has advantages, including increasing eye health, boosting muscle power and endurance, and shielding the skin from UVA damage, inflammation, and early aging. Animals need it for various purposes, such as immune system functions and regeneration. It is a potent coloring agent. Other carotenoids are catechin and phycocyanobilin (**Figure 3**).

4.2 Phycobilin pigments

Microalgae form accessory pigments like phycobiliproteins. These pigments improve the light energy utilization efficiency of algae and protect it from solar

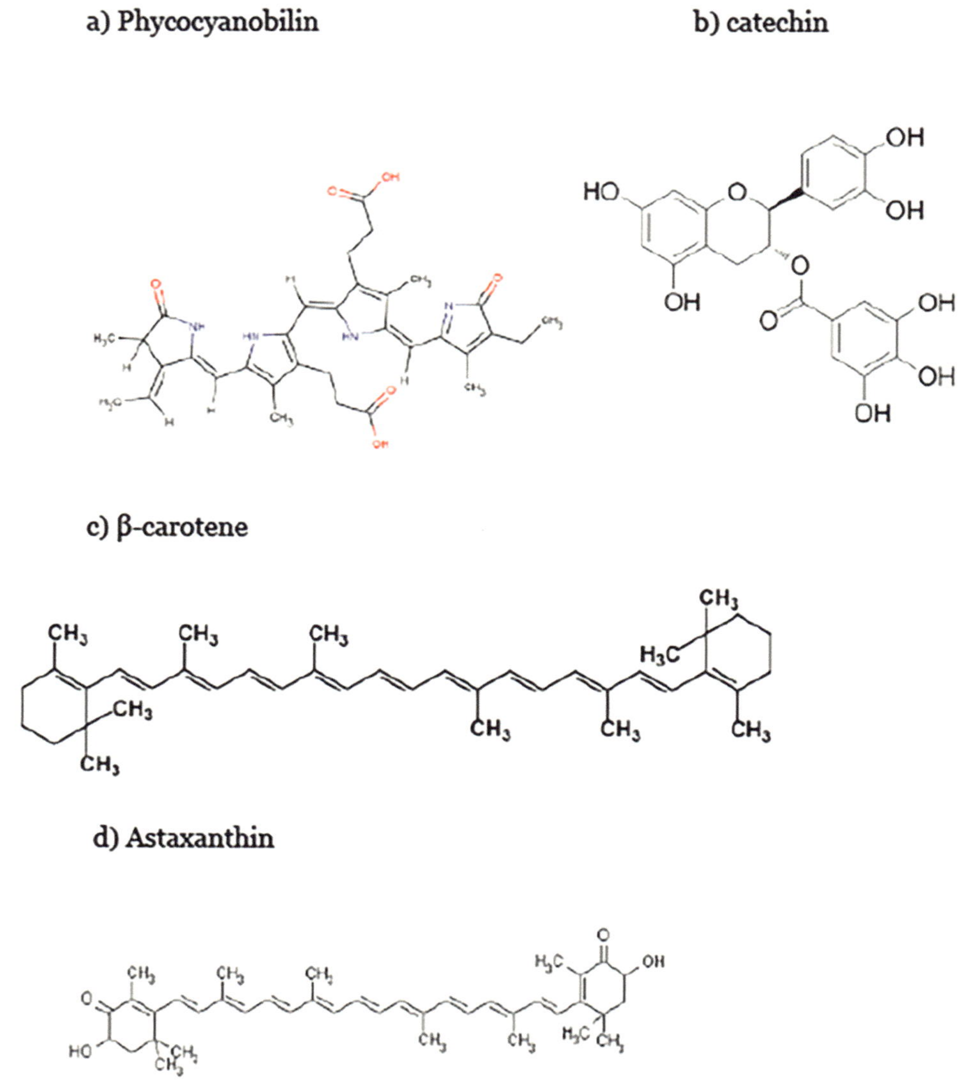

Figure 3.
Chemical structure of some carotenoids. a) Phycocyanobilin b) Catechin, c) β-carotene d) Astaxanthin.

radiation and its effects. They are antioxidants in feed and humans. Phycobiliproteins are the major component of light-harvesting antenna pigments called phycobilisomes. They are present in Rhodophyta (red algae), Cryptomonads algae, and Cyanobacteria (blue-green algae) [71].

The cyanobacterium *Spirulina* (Arthrospira), which produces phycocyanin (blue), and the rhodophyte porphyridium, which produces phycoerythrin(red), are the main sources of phycobiliproteins. In Japan and China, phycocyanin is utilized in chewing gum, candies, dairy goods, jellies, ice cream, soft beverages, as well as in cosmetics like lipsticks. Phycocyanin is a versatile blue coloring agent that gives jelly and confections a vibrant blue color [71].

4.3 Phenolic and polyphenols compounds

The secondary metabolites are polyphenols. They are a collection of chemical substances found in aquatic macrophytes and terrestrial plants. Phenolic compounds are present in edible plants, and their structure contains benzene [71]. Plants frequently contain phenolic chemicals. Polyphenols contain tannins, phenolic acids, flavonoids, tocopherols, and lignin.

Class	Metabolites	Species	Reference
1. Carotenoids	Asctaxanthin	*Chlorella zofingiensis, Chlorococcum* sp., *Haematococcus pluvialis, Scenedesmus* sp.	[71, 72]
	β-carotene	*Dunaliella bardawil, Dunaliella salina, Dunaliella tertiolecta, Scenedesmus almeriensis*	
	Canthaxanthin	*Coelastrella striolata, C. zofingiensis, D. salina, Scenedesmus komareckii*	
	Echinenone	*Botryococcus braunii*	
	Fucoxanthin	*Isochrysis galbana, Phaeodactylu tricornutum*	
	Lutein	*Chlorella protothecoides, Chlorella zofingiensis, Chlorococcum citriforme, Muriellopsis sp., S. almeriensis*	
	Lycopene	*Chlorella ellipsoidea, Chlorella marina, D. tertiolecta*	
	Peridinin	*Amphidinium carterae*	
	Phytoene	*Dunaliella* sp.	
	Phytofluene	*Dunaliella* sp.	
2. Polysaccharides	Crude polysaccharide extracts	*Chlorella stigmatophora, P. tricornutum, P. cruentum, Rhodella reticulata*	[73]
3. Phycobiliproteins	(A) Phycoerythrin – Red (B) Phycoerythrin – Blue	*Arthrospira platensis, Limnothrix* sp., *Nostoc* sp., *Phorphyridium aerugineum, Phormidium ceylanicum, Synechococcus lividus*	[74]
4. Polyphenols	(A) Phenolic acids, (B) Flavonoids-Marennin	*Ankistrodesmus* sp., *A. platensis, Caespitella pascheri, Euglena cantabrica, Leptolyngbya protospira, Nostoc commune, Nodularia spumigena, Phormidiochaete* sp., *Spirogyra* sp.	[75]

Table 3.
Algal metabolites.

4.4 Sulfated polysaccharides

There are substances called polysaccharides in plants, animals, algae, microbes, and other natural products. They comprise numerous monosaccharides connected by various glycosidic linkages and contain polymeric carbohydrate structures. Sulfated polysaccharides are of non-animal origin and are most abundant. Most sulfated polysaccharides found in nature are complex combinations of molecules with a wide range of structures and activity [71]. They frequently occur in nature. Sulfated polysaccharides with a variety of biological functions are primarily found in seaweed. Fucoidan is a complex sulfated polysaccharide that is present mainly in the cell wall fluid of several species. It contains L-fructose and sulfate, ester groups.

Some of the important metabolites found in different algal species are listed in **Table 3**.

5. Applications

5.1 Potential uses for antioxidants

Cyanobacteria's ancestors produced the first biogenic molecular oxygen on Earth, but they are still unknown how they handled oxidative stress [71]. We explore the advancement of superoxide dismutase proteins (Turfs) equipped for eliminating superoxide-free revolutionaries and gauge the beginning of Cyanobacteria. Our microfossil-adjusted Bayesian atomic timekeepers foresee that stem cyanobacteria emerged quite a while back. The development of NiSOD is especially captivating because it concurs with the intrusion of the vast sea by cyanobacteria [21]. Microalgal biotechnology can expand into regions and climates that are unfavorable for agriculture, such as deserts and seashores, and can reach higher productivity. Additionally, aquaculture and life-support systems depend on microalgae cultivation as feed, and they effectively remove nutrients from water [70].

5.2 Potential application in the agricultural sector

Cyanobacteria have been studied thoroughly and have established a solid ground in the agricultural sector. Utilizing microalgae and cyanobacteria to increase agricultural productivity sustainably and efficiently has various potential advantages [76]. A significant source of a wide range of bioactive substances that can control numerous plant response processes is microalgae and cyanobacteria: the enhancement of soil fertility and plant nutrition; the defense of plants against factors both biotic and abiotic; and promotion of growth. Hence, we can conclude that using microalgal/cyanobacterial biomass (or their extracts) instead of chemical-based fertilizers, insecticides, and growth promoters can be an extremely feasible and sustainable option in the agricultural sector. Additionally, the utilization of these biologically based organisms gives rise to a significant step in improving agricultural productivity, which is crucial to achieving the ever-increasing objectives for food items, pharmaceutical items, toxic items, and antitoxic items which are heavily mandated by growing global population [71].

5.3 Potential application in pharmaceuticals and cosmetics sector

Cyanobacteria are a rich source of organic chemicals and can be utilized to make food. Cyanobacteria-derived compounds are often used in cosmetics as thickening,

water-binding, and antioxidant ingredients. Cosmetic companies typically base their skin or health claims on ingredients such as carrageenan, vitamin A, polysaccharides, iron, phosphorus, salt, copper, vitamin B1, and minerals such as calcium, magnesium, or others [71].

Since they frequently target tubulin or actin filaments in eukaryotic cells, cyanobacteria cytotoxic metabolites are appealing anticancer medicines. Dolastatins, produced by NRPS-PKS enzymes and present in *Leptolyngbya* and *Simploca* sp., can interfere with the development of microtubules. Other cyanobacterial byproducts, such as the cyclic depsipeptide derivatives known as lyngbyastatins, which are also suspected elastase inhibitors, function as protease inhibitors [70]. A mixture NRPS-PKS pathway is utilized in the biosynthesis of apratoxins such apratoxin-a from *Lyngbya majuscule*. Due to its capacity to cause G1-phase cell cycle arrest and death, it is cytotoxic.

In the cosmetics sector, secondary metabolites can be employed as natural components. Sunscreens contain photoprotective MAAs that protect the skin from UVR damage. In addition to serving as natural colorants, pigments like carotenoids and phycobiliproteins can act as antioxidants to shield the skin from UV-induced mutilation [71].

6. Future prospects

Due to their intricate structures and a range of bioactivities, these secondary metabolites can also be used as lead molecules in developing new drugs. A biosynthetic pathway study employing genomic data with over 208 publicly available cyanobacterial genome sequences can find new natural compounds. Even though cyanobacterial secondary metabolites have been the subject of intensive research, a variety of species still need to be sequenced and examined, and there are still a lot of secondary metabolites that may be significant but have not yet been identified. The growth study of cyanobacteria under peroxidation is still in its initial phases. Most of the information we know about the production of reactive oxygen species in the photosystem mechanism comes from plant findings. Still, cyanobacteria, as a distinct type for photochemical research, will help us understand peroxidation in photosynthesis in general. Analyzing the particular reaction for each reactive form is challenging because many ROS are created concurrently in cells. Future studies should concentrate on the biological consequences and particular targets of specific ROS species in cyanobacteria and the functional reactions they cause. In recent years, significant advancements have been achieved in identifying ROS-scavenging enzymes. The recognition or finding of ROS-scavenging enzymes has advanced significantly in recent years. Still, much more investigation is required to comprehend their function in vivo fully, as well as to determine which areas of the cell they act in and the range of oxidants they can detoxify.

7. Conclusion

Natural algae antioxidants are significant bioactive substances that help fight various ailments and shield cells from oxidative stress. It is an important source of chemicals that are neuroprotective. Algae have exceptional nutritional value, and therapeutic properties have higher demands for natural algal products. This gives

microalgae clear benefits over conventional components, making them worth investigating for future use in the feed, food, cosmetic, and pharmaceutical industries. The extensive evolutionary history of cyanobacteria has resulted in adaptations that allow them to cope with natural and man-made stress. The enormous quantity of secondary metabolites produced by cyanobacteria, each with its unique roles supporting the organism's survival, results from the diversity of their morphological, biochemical, and physiological makeup.

Conflict of interest

The authors declare no conflict of interest.

Author details

Rashi Tyagi, Pankaj Kumar Singh and Archana Tiwari*
Diatom Research Laboratory, Amity Institute of Biotechnology, Amity University, Noida, Uttar Pradesh, India

*Address all correspondence to: panarchana@gmail.com

References

[1] Cervantes, Gracia K, Llanas-Cornejo D, Husi H. CVD and oxidative stress. Journal of Clinical Medicine. 2017;**6**(2):22. DOI: 10.3390/jcm6020022

[2] Talero E, García-Mauriño S, Ávila-Román J, Rodríguez-Luna A, Alcaide A, Motilva V. Bioactive compounds isolated from microalgae in chronic inflammation and Cancer. Marine Drugs. 2015;**13**(10):6152-6209. DOI: 10.3390/md13106152

[3] Meireles dos Santos A, Vieira KR, Basso Sartori R, Meireles dos Santos A, Queiroz MI, Queiroz Zepka L, et al. Heterotrophic cultivation of cyanobacteria: Study of the effect of exogenous sources of organic carbon, an absolute amount of nutrients, and stirring speed on biomass and lipid productivity. Frontiers in Bioengineering and Biotechnology. 2017;**5**:1-7. DOI: 10.3389/fbioe.2017.00012

[4] Chong S, Chen C, Ge H, Xie XS. Mechanism of transcriptional bursting in bacteria. Cell. 2014;**158**(2):314-326. DOI: 10.1016/j.cell.2014.05.038

[5] Obinger C, Regelsberger G, Furtmüller PG, Jakopitsch C, Rüker F, Pircher A, et al. Catalase-peroxidases in cyanobacteria--similarities and differences to ascorbate peroxidases. Free Radical Research. 1999;**31**(Suppl):S243-S249. DOI: 10.1080/10715769900301571

[6] Volk RB. A newly developed assay for the quantitative determination of antimicrobial (anticyanobacterial) activity of both hydrophilic and lipophilic test compounds without any restriction. Microbiological Research. 2008;**163**(2):161-167. DOI: 10.1016/j.micres.2006.03.015

[7] Rodrigues JLM, Silva-Stenico ME, Gomes JE, Lopes JRS, Tsai SM. Detection and diversity assessment of Xylella fastidiosa in field-collected plant and insect samples by using 16S rRNA and gyrB sequences. Applied and Environmental Microbiology. 2003;**69**(7):4249-4255. DOI: 10.1128/aem.69.7.4249-4255.2

[8] Krishnamurthy P, Wadhwani A. Antioxidant enzymes and human health. In: Antioxidant Enzyme. London, UK: Intech Open; 2012. DOI: 10.5772/48109

[9] Singh JS, Kumar A, Rai AN, Singh DP. Cyanobacteria: A precious bio-resource in agriculture, ecosystem, and environmental sustainability. Frontiers in Microbiology. 2016;**2016**:7. DOI: 10.3389/fmicb.2016.00529

[10] Hossain MF, Ratnayake RR, Meerajini K, Wasantha Kumara KL. Antioxidant properties in some selected cyanobacteria isolated from freshwater bodies of Sri Lanka. Food Science & Nutrition. 2016;**4**(5):753-758. DOI: 10.1002/fsn3.340

[11] Huang KH, Ravindran V, Li X, Ravindran G, Bryden WL. Apparent ileal digestibility of amino acids in feed ingredients determined with broilers and layers. Journal of the Science of Food and Agriculture. 2007;**87**(1):47-53. DOI: 10.1002/jsfa.2667

[12] Lv S, Cheng J, Sun A, Li J, Wang W, Guan G, et al. Mesenchymal stem cells transplantation ameliorates glomerular injury in streptozotocin-induced diabetic nephropathy in rats via inhibiting oxidative stress. Diabetes Research and Clinical Practice. 2014;**104**(1):143-154. DOI: 10.1016/j.diabres.2014.01.011

[13] Ramos J, Ariño J, Sychrová H. Alkali-metal-cation influx and efflux systems in nonconventional yeast species. FEMS Microbiology Letters. 2011;**317**(1):1-8. DOI: 10.1111/j.1574-6968.2011.02214. x

[14] Montero-LobatoZ,Ramos-MerchanteA, FuentesJ,SayagoA,Fernández-RecamalesÁ, Martínez-Espinosa R, et al. Optimization of growth and carotenoid production by *Haloferax mediterranei* using response surface methodology. Marine Drugs. 2018;**16**(10):372

[15] Gong M, Bassi A. Carotenoids from microalgae: A review of recent developments. Biotechnology Advances. 2016;**34**(8):1396-1412. DOI: 10.1016/j.biotechadv.2016.10.005

[16] Yu C, Liu Y, Ma T, Liu K, Xu S, Zhang Y, et al. Small molecules enhance CRISPR genome editing in pluripotent stem cells. Cell Stem Cell. 2015;**16**(2):142-147. DOI: 10.1016/j.stem.2015.01.003

[17] Varshney GK, Pei W, LaFave MC, Idol J, Xu L, Gallardo V, et al. High-throughput gene targeting and phenotyping in zebrafish using CRISPR/Cas9. Genome Research. 2015;**25**(7):1030-1042. DOI: 10.1101/gr.186379.114

[18] Sun JC, Beilke JN, Lanier LL. Adaptive immune features of natural killer cells. Nature. 2009;**457**(7229):557-561. DOI: 10.1038/nature07665

[19] Chen C, Huang J, Hsiao Y. Knowledge management and innovativeness. International Journal of Manpower. 2010;**31**(8):848-870. DOI: 10.1108/01437721011088548

[20] Banerjee M, Raghavan PS, Ballal A, Rajaram H, Apte SK. Oxidative stress management in the filamentous, heterocystous, diazotrophic cyanobacterium, Anabaena PCC7120. Photosynthesis Research. 2013;**118**:59-70. DOI: 10.1007/s11120-013-9929-8

[21] Boden JS, Konhauser KO, Robbins LJ, et al. Timing the evolution of antioxidant enzymes in cyanobacteria. Nature Communications. 2021;**12**:4742. DOI: 10.1038/s41467-021-24396-y

[22] Zhang J, Liu L, Ren Y, Chen F. Characterization of exopolysaccharides produced by microalgae with antitumor activity on human colon cancer cells. International Journal of Biological Macromolecules. 2009;**128**:761-767. DOI: 10.1016/j.ijbiomac.2019.02.009

[23] Singh PK, Bhattacharjya R, Saxena A, Thakur IS, Tiwari A. Envisaging the role of pharmaceutical contaminant 17-β estradiol on growth and lipid productivity of marine diatom *Chaetoceros gracilis*. Bioresource Technology. 2022a;**346**:126642. DOI: 10.1016/J.BIORTECH.2021.126642

[24] Cha KH, Koo SY, Lee DU. Antiproliferative effects of carotenoids extracted from *Chlorella ellipsoidea* and *Chlorella vulgaris* on human colon cancer cells. Journal of Agricultural and Food Chemistry. 2008;**56**(22):10521-10526. DOI: 10.1021/jf802111x

[25] Foo SC, Yusoff FM, Imam MU, Foo JB, Ismail N, Azmi NH, et al. Increased fucoxanthin in *Chatoceros calcitrans* extract exacerbates apoptosis in liver cancer cells via multiple targeted cellular pathways. Biotechnological Reports. 2018;**20**:e00296. DOI: 10.1016/j.btre.2018.e00296

[26] Ganesan P, Noda K, Manabe Y, Ohkubo T, Tanaka Y, Maoka T, et al. Siphonaxanthin, a marine carotenoid from green algae, effectively induces apoptosis in human leukemia (HL-60) cells. Biochimica et Biophysica Acta. 2011;**1810**(5):497-503. DOI: 10.1016/j.bbagen.2011.02.008

[27] Baudelet PH, Gagez AL, Bérard JB, Juin C, Bridiau N, Kaas R, et al. Antiproliferative activity of *Cyanophora paradoxa* pigments in melanoma, breast and lung cancer cells. Marine Drugs. 2013;**11**(11):4390-4406. DOI: 10.3390/md11114390

[28] Jayappriyan KR, Rajkumar R, Venkatakrishnan V, Nagaraj S, Rengasamy R. In vitro anticancer activity of natural β-carotene from *Dunaliella salina* EU5891199 in PC-3 cells. Biomedicine & Preventive Nutrition. 2013;**3**(2):99-105. DOI: 10.1016/j.bionut.2012.08.003

[29] Ferdous UT, Yusof ZNB. Medicinal prospects of antioxidants from algal sources in Cancer therapy. Frontiers in Phfromcology. 2021;**12**:593116. DOI: 10.3389/fphar.2021.593116

[30] Pasquet V, Morisset P, Ihammouine S, Chepied A, Aumailley L, Berard JB, et al. Antiproliferative activity of violaxanthin isolated from bioguided fractionation of *Dunaliella tertiolecta* extracts. Marine Drugs. 2011;**9**(5):819-831. DOI: 10.3390/md9050819

[31] Nagaraj S, Rajaram MG, Arulmurugan P, Baskaraboopathy A, Karuppasamy K, Jayappriyan KR. Antiproliferative potential of astaxanthin rich alga *Haematococcus pluvialis* flotow on human hepatic cancer (HepG2) cell line. Biomedicine & Preventive Nutrition. 2012;**2**(3):149-153. DOI: 10.1016/j.bionut.2012.03.009

[32] Sanjeewa KKA, Fernando IPS, Samarakoon KW, Lakmal HHC, Kim EA, Kwon ON. Anti-inflammatory and anti-cancer activities of sterol rich fraction of cultured marine microalga *Nannochloropsis oculata*. Algae. 2016;**31**(3):277-287. DOI: 10.4490/algae.2016.31.6.29

[33] Sayegh F, Elazzazy A, Bellou S, Moustogianni A, Elkady AI, Baeshen MN. Production of polyunsaturated single cell oils possessing antimicrobial and anticancer properties. Annales de Microbiologie. 2016;**66**(3):937-948. DOI: 10.1007/s13213-015-1176-0

[34] Bhattacharjya R, Marella TK, Tiwari A, Saxena A, Singh PK, Mishra B. Bioprospecting of marine diatoms *Thalassiosira*, *Skeletonema* and *Chaetoceros* for lipids and other value-added products. Bioresource Technology. 2020;**2020**:124073. DOI: 10.1016/j.biortech.2020.1240

[35] Singh PK, Bhattacharjya R, Saxena A, Mishra B, Tiwari A. Utilization of wastewater as nutrient media and biomass valorization in marine Chrysophytes- *Chaetoceros* and *Isochrysis*. Energy Conversion and Management: X. 2021;**10**:100062. DOI: 10.1016/j.ecmx.2020.100062

[36] Singh PK, Bhattacharjya R, Mishra B, Saxena A, Tiwari A. A multifaceted approach towards valorizing diatom *Thalassiosira weissflogii*, cultivated on diluted municipal wastewater for enhanced biodiesel production. Fuel. 2022b;**328**:125311. DOI: 10.1016/J.FUEL.2022.125311

[37] Yang S, Wan H, Wang R, Hao D. Sulfated polysaccharides from *Phaeodactylum tricornutum*: Isolation, structural characteristics, and inhibiting HepG2 growth activity in vitro. PeerJ. 2019;**7**:e6409. DOI: 10.7717/peerj.6409

[38] Juin C, de Oliveira Junior RG, Fleury A, Oudinet C, Pytowski L, Bérard JB. Zeaxanthin from *Porphyridium purpureum* induces apoptosis in human melanoma cells expressing the oncogenic BRAF V600E mutation and sensitizes them to the BRAF inhibitor vemurafenib. Brazilian Journal of Pharmacognosy. 2018;**28**(4):457-467. DOI: 10.1016/j.bjp.2018.05.009

[39] Bergea JP, Debiton E, Dumay J, Durand P, Barthomeuf C. In vitro anti-inflammatory and anti-proliferative activity of sulfolipids from the red alga *Porphyridium cruentum*. Journal of Agricultural Food Chemistry. 2002;**50**:6227-6232. DOI: 10.1021/jf020290y

[40] Jiang L, Wang Y, Liu G, Liu H, Zhu F, Ji H. C-Phycocyanin exerts anti-cancer effects via the MAPK signaling pathway in MDA-MB-231 cells. Cancer Cellular International. 2018;**18**(1):12. DOI: 10.1186/s12935-018-0511-5

[41] Chen X, Song L, Wang H, Liu S, Yu H, Wang X. Partial characterization, the immune modulation and anticancer activities of sulphated polysaccharides from filamentous microalgae *Tribonema* sp. Molecules. 2019;**24**(2):322. DOI: 10.3390/molecules24020322

[42] Miyake C, Michihata F, Asada K. Scavenging of hydrogen peroxide in prokaryotic and eukaryotic algae: Acquisition of ascorbate peroxidase during the evolution of cyanobacteria. Plant and Cell Physiology. 1991;**32**:33-43

[43] Bernroitner M, Zamocky M, Furtmüller PG, Peschek GA, Obinger C. Occurrence, phylogeny, structure, and function of catalases and peroxidases in cyanobacteria. Journal of Experimental Botany. 2009;**60**(2):423-440. DOI: 10.1093/jxb/ern309

[44] RegelsbergerG AW, Rüker F, Peschek GA, Jakopitsch C, Paumann M, Obinger C. Biochemical characterization of a membrane-bound manganese-containing superoxide dismutase from the cyanobacterium *Anabaena* PCC 7120. Journal of Biological Chemistry. 2002;**277**(46):43615-43622. DOI: 10.1074/jbc.m207691200

[45] Priya B, Sivaprasanth RK, Jensi VD, et al. Characterization of manganese superoxide dismutase from a marine cyanobacterium *Leptolyngbya valderiana* BDU20041. Saline Systems. 2010;**6**:6. DOI: 10.1186/1746-1448-6-6

[46] Sheng Y, Abreu I, Cabelli DE, et al. Superoxide dismutases and superoxide reductases. Chemical Reviews. 2014;**2014**(114):3854-3918. DOI: 10.1021/cr4005296

[47] Priya B, Premanandh J, Dhanalakshmi RT, et al. Comparative analysis of cyanobacterial superoxide dismutases to discriminate canonical forms. BMC Genomics. 2007;**8**:435. DOI: 10.1186/1471-2164-8-435

[48] Asada K. The water-water cycle in chloroplasts: Scavenging of active oxygens and dissipation of excess photons. Annual Review of Plant Physiology and Plant Molecular Biology. 1999;**50**:601-639

[49] Wood ZA, Schröder E, Robin Harris J, Poole LB. Structure, mechanism and regulation of peroxiredoxins. Trends in Biochemical Science. 2000;**28**:32-40

[50] Kumar A, Ramamoorthy D, Verma DK, Kumar A, Kumar N, Kanak KR, et al. Antioxidant and phytonutrient activities of Spirulina platensis. Energy Nexus. 2022;**2022**:100070. DOI: 10.1016/j.nexus.2022.100070

[51] Bhagavathy S, Sumathi P, Sherene J, Bell I. Green algae *Chlorococcum humicola*- a new source of bioactive compounds with antimicrobial activity. Asian Pac. Journal of Tropical Biomedicine. 2011;**1**:S1. DOI: 10.1016/S2221-1691(11)60111-1

[52] Paige Stout E, Kubanek J. 2.03 - Marine macroalgal natural products. In: Liu H-W, Mander L, editors. Comprehensive Natural Products II. Elsevier; 2010. pp. 41-65

[53] Giorgos M, Elias M. Microalgae for high-value compounds and biofuels production: A review with focus on cultivation under stress conditions. Biotechnology Advances. 2013;**31**(8):1532-1542. DOI: 10.1016/j.biotechadv.2013.07.011

[54] Saini DK, Pabbi S, Shukla P. Cyanobacterial pigments: Perspectives and biotechnological approaches. Food and Chemical Toxicology. 2018;**120**:616-624. DOI: 10.1016/j.fct.2018.08.002

[55] Singh D, Prabha R, Meena K, Sharma L, Sharma A. Induced accumulation of polyphenolics and flavonoids in Cyanobacteria under salt stress protects organisms through enhanced antioxidant activity. American Journal of Plant Sciences. 2014;**5**:726-735. DOI: 10.4236/ajps.2014.55087

[56] Parys S, Kehraus S, Krick A, Glombitza KW, Carmeli S, Klimo K, et al. In vitro chemopreventive potential of fucophlorethols from the brown alga *Fucus vesiculosus L* by antioxidant activity and inhibition of selected cytochrome P450 enzymes. Phytochemistry. 2010;**71**(2-3):221-229. DOI: 10.1016/j.phytochem.2009.10.020

[57] Kar S, Sharma G, Das PK. Fucoidan cures infection with both antimony-susceptible and-resistant strains of Leishmania donovani through Th1 response and macrophage-derived oxidants. The Journal of Antimicrobial Chemotherapy. 2011;**66**:618-625. DOI: 10.1093/jac/dkq502

[58] Anisha GS, Padmakumari S, Patel AK, Pandey A, Singhania RR. Fucoidan from marine macroalgae: Biological actions and applications in regenerative medicine, drug delivery systems and food industry. Bioengineering. 2022;**9**:472. DOI: 10.3390/bioengineering9090472

[59] Lee AH, Shin HY, Park JH, et al. Fucoxanthin from microalgae *Phaeodactylum tricornutum* inhibits pro-inflammatory cytokines by regulating both NF-κB and NLRP3 inflammasome activation. Scientific Reports. 2023;**11**:543. DOI: 10.1038/s41598-020-80748-6

[60] Fenoradosoa DC, Andrée T, Philippe M. Galactans: An overview of their most important sourcing and applications as natural polysaccharides. Brazilian Archives of Biology and Technology. 2011;**54**(6):1075-1092. DOI: 10.1590/S1516-89132011000600002

[61] Gouda M, Tadda MA, Zhao Y, Farmanullah F, Chu B, Li X, et al. Microalgae bioactive carbohydrates as a novel sustainable and eco-friendly source of prebiotics: Emerging health functionality and recent Technologies for Extraction and Detection. Frontiers in Nutrition. 2022;**9**:806692. DOI: 10.3389/fnut.2022.806692

[62] Dang TT, Van Vuong Q, Schreider MJ, Bowyer MC, Van Altena IA, Scarlett CJ. Optimisation of ultrasound-assisted extraction conditions for phenolic content and antioxidant activities of the alga *Hormosira banksii* using response surface methodology. Journal of Applied Phycology. 2017;**29**:3161-3173. DOI: 10.1007/s10811-017-1162-y

[63] Haoujar I, Cacciola F, Abrini J, Mangraviti D, Giuffrida D, Oulad el Majdoub Y, et al. The contribution of carotenoids, phenolic compounds, and flavonoids to the antioxidative properties of marine microalgae isolated from Mediterranean Morocco. Molecules. 2019;**24**(22):4037. DOI: 10.3390/molecules24224037

[64] Glazer A. Phycobiliproteins — A family of valuable, widely used

fluorophores. Journal of Applied Phycology. 1994;**6**:105-112

[65] Wei Y, Liu Q, Xu C, Yu J, Zhao L, Guo Q. Damage to the membrane permeability and cell death of Vibrio parahaemolyticus caused by phlorotannins with low molecular weight from Sargassum thunbergii. Journal of Aquatic Food Product Technology. 2016;**25**:323-333. DOI: 10.1080/10498850.2013.851757

[66] Mittal R, Tavanandi HA, Mantri VA, Raghavarao KSMS. Ultrasound-assisted methods for enhanced extraction of phycobiliproteins from marine macro-algae, *Gelidium pusillum* (Rhodophyta). Ultrasonics Sonochemistry. 2017;**38**:92-103. DOI: 10.1016/j.ultsonch.2017.02.030

[67] Hirschberg J, Chamovitz D. Carotenoids in Cyanobacteria. The Molecular Biology of Cyanobacteria. 1994;**1994**:559-579. DOI: 10.1007/978-94-011-0227-8_18

[68] Zanchett G, Oliveira-Filho EC. Cyanobacteria and cyanotoxins: From impacts on aquatic ecosystems and human health to anticarcinogenic effects. Toxins (Basel). 2013;**5**(10):1896-1917. DOI: 10.3390/toxins5101896

[69] Tanaka NK, Endo K, Ito K. Organization of antennal lobe-associated neurons in adult Drosophila melanogaster brain. The Journal of Comparative Neurology. 2012;**520**(18):4067-4130

[70] Capone DG, Burns JA, Montoya JP, Subramaniam A, Mahaffey AC, Gunderson T, et al. Nitrogen fixation by *Trichodesmium* spp.: An important source of new nitrogen to the tropical and subtropical North Atlantic Ocean. Global Biogeochem Cycles. 2005;**19**:GB2024. DOI: 10.1029/2004GB002331

[71] Hassan S, Meenatchi R, Pachillu K, Bansal S, Brindangnanam P, Arockiaraj J, et al. Identification and characterization of the novel bioactive compounds from microalgae and cyanobacteria for pharmaceutical and nutraceutical applications. Journal of Basic Microbiology. 2022;**62**:999-1029. DOI: 10.1002/jobm.202100477

[72] Novoveská L, Ross ME, Stanley MS, Pradelles R, Wasiolek V, Sassi JF. Microalgal carotenoids: A review of production, current markets, regulations, and future direction. Marine Drugs. 2019;**17**:640

[73] Sathasivam R, Radhakrishnan R, Hashem A, Abd-Allah EF. Microalgae metabolites: A rich source for food and medicine. Saudi Journal of Biological Science. 2019;**26**:709-722

[74] Manirafasha E, Ndikubwimana T, Zeng X, Lu Y, Jing K. Phycobiliprotein: Potential microalgae derived pharmaceutical and biological reagent. Biochemical Engineering Journal. 2016;**109**:282-296

[75] Levasseur W, Perré P, Pozzobon V. A review of high value-added molecules production by microalgae in light of the classification. Biotechnology Advances. 2020;**41**:107545

[76] Abed RMM, Dobretsov S, Sudesh K. Applications of cyanobacteria in biotechnology. Journal of Applied Microbiology. 2009;**106**(1):1-12. DOI: 10.1111/j.1365-2672.2008. 03918.x

Chapter 4

Perspective Chapter: Cyanobacteria – A Futuristic Effective Tool in Sustainable Agriculture

Eman Elagamey, Magdi A.E. Abdellatef and Hassan E. Flefel

Abstract

Cyanobacteria are bioactive photosynthetic prokaryotes that have a superior ability to fix atmospheric nitrogen and are highly competitive in the microflora community. They also improve the physical and chemical properties of the soil and increase its water-holding capacity. Therefore, cyanobacteria are used as biofertilizers in agriculture. Cyanobacteria are able to promote plant growth by providing nutrients and producing many highly effective chemical compounds, such as enzymes and hormones, in the plant rhizosphere, giving the plant a highly competitive ability. In addition to activating plant defense responses against soil-borne pathogens, they have an effective strategy as a biocide against bacteria, fungi, and nematodes that attack plants. With multiple beneficial biological roles, the environmentally friendly cyanobacteria occupied the role of the maestro in sustainable agriculture.

Keywords: cyanobacteria, sustainable agriculture, biofertilizer, nitrogen fixation, abiotic stress, antimicrobial activity

1. Introduction

Given the ongoing increase in the world's population and the depletion of food resources, our society currently needs a sustainable supply of agricultural productivity that poses no environmental risks [1]. Plants are constantly affected by abiotic stresses, such as drought, salinity, cold, heat, and nutrient deficiencies, as well as biotic stress, including pathogens and pests. In addition to climatic changes that greatly affect soil fertility, virulence of pests and diseases, and plant-producing biomass and seeds [2]. In nature, the interaction continues between biotic stress and plants, causing dynamic changes in their activities and composition under changing environmental conditions. Beneficial microorganisms play an effective role in maintaining the balance of this interaction in a way that is in the interest of the plant at the expense of biotic stress. Furthermore, plants can more effectively withstand abiotic stress, enhance nutrient uptake and utilization, and increase photosynthetic activity by virtue of the mechanisms carried out by beneficial microorganisms, which leads to higher yield [3].

Beneficial microorganisms can act as biopesticides by attacking phytopathogens directly and limiting their population by competition for space, nutrients, and the production of antimicrobial compounds [4]. Furthermore, beneficial microorganisms can induce plants to pre-activate the defensive responses controlled by plant hormones in order to combat infections more rapidly and successfully. This is referred to as systemic acquired resistance [5]. Among beneficial microorganisms are cyanobacteria. Cyanobacteria are photosynthetic prokaryotic organisms, extremely varied groups that can be found in practically all of the world's ecosystems.

Cyanobacteria occur in unicellular, colonial, or multicellular filamentous forms. They are considered a subset of the bacterial kingdom. This subset is responsible for a significant amount of N_2 fixation, reduction of the level of CO_2, solubilization of phosphate, and the production of plant growth regulators by releasing phytohormones, polypeptides, amino acids, polysaccharides, and siderophores [6]. Cyanobacteria are composed of numerous organic inclusion units capable of carrying out a wide range of specialized functions, which give cyanobacteria their unique tasks and applications in sustainable agriculture [7]. The components that make up the structure of cyanobacteria are light-harvesting antennae, phycobilisomes, polyphosphate bodies, cyanophycin granules, polyhydroxyalkanoate granules, carboxysomes, lipid bodies, thylakoids, DNA-containing areas, and ribosomes [8] (**Figure 1**). Cyanobacteria have chlorophyll-a, which engages it in oxygenic photosynthesis, carotenoids that protect chlorophyll-a from oxidative degradation, and specific pigments called phycobilins that are bound to water-soluble proteins [9]. Flagella are not present in cyanobacteria [10].

Some kinds of cyanobacteria contain specialized cells called heterocytes and akinetes that are morphologically distinct from vegetative cells. The position, amount, and distribution of heterocytes and akinetes are significant morphological characteristics of cyanobacteria species and genera. Heterocytes are specialized cells that allow nitrogenase to fix atmospheric nitrogen by reducing it to ammonium, a process known as diazotrophy [11]. Akinetes contain granules of glycogen and cyanophycin but no polyphosphate granules and have a multilayered cell wall [9].

2. Characterization of cyanobacteria

Cyanobacteria are distinguished from most other microalgae by their lack of a cell nucleus and other cell organelles. They lack chloroplasts and have instead simple thylakoids, which are the location of the light-dependent processes necessary for

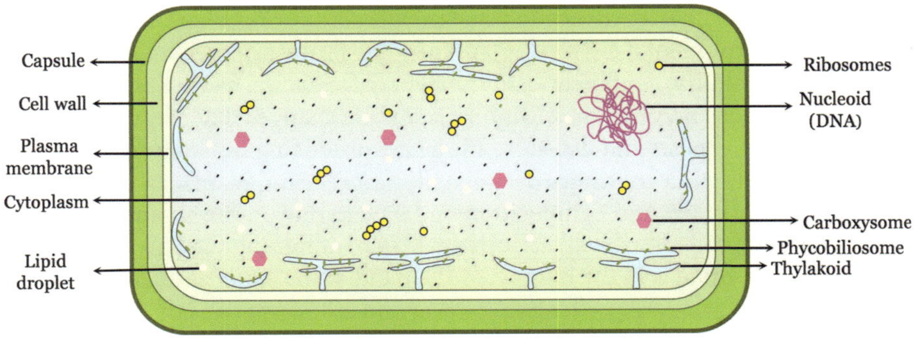

Figure 1.
Cyanobacterial cell structure.

photosynthesis. Cyanobacteria exhibit a variety of traits that can be utilized for microscopic analysis and identification, including the size and form of the cells, the presence of subcellular structures, and the presence of specialized cells. Flagella, which are present in many other bacterial or phytoplankton taxa, are absent in cyanobacteria. However, many cyanobacteria, especially filamentous varieties, exhibit gliding movement. Cyanobacteria have not been shown to reproduce sexually. The division of vegetative cells is their unique asexual method of reproduction. Cyanobacterial cells can be spherical, cylindrical, barrel-shaped, ellipsoid, conical, or disc-shaped. The critical abiotic parameters that determine the success of cyanobacterial growth are light, pH, temperature, water, CO_2, and nutrient supplements [9, 12].

3. Symbiotic association between plant and cyanobacteria

In a symbiotic relationship, both organisms can benefit from each other in various ways. The filamentous cyanobacteria live in symbiosis with a wide range of eukaryotic hosts, including plants and fungi [13]. The cyanobacteria that will form symbiotic relationships with the plants are called cyanobionts, which can grow inside the host or more or less firmly attach themselves to the host [14]. The plants provide cyanobacteria with carbon sources, for example, sucrose. Cyanobacteria have the ability to fix nitrogen from the air in heterocysts, which benefits plants by supplying them with nitrogen (**Figure 2**). Therefore, the symbiotic cyanobacteria are mostly heterocyst-forming strains. They are virtually entirely associated with the genera Nostoc and Anabaena [15]. Cyanobacteria provide plants with about 88% of the fixed nitrogen in the form of NH_3 and keep only 12% for themselves [16]. In addition to a symbiotic relationship between cyanobacteria and whole plants, there is also a symbiotic relationship between cyanobacteria and plant tissues. Cyanobacteria were found to have colonized different areas of wheat, where they were abundantly present around the root, in the spaces between root epidermal cells and cortex, and as single cells within the stem or on the surface of leaves [17]. Due to the symbiotic relationship between Gunnera and Nostoc, the number of heterocysts formed increased by up to 80%. This is evidence that the symbiotic relationship has different effects on the growth and development of cyanobacteria. Leghemoglobin concentration in chickpea root nodules increases as a result of the simultaneous inoculation of the

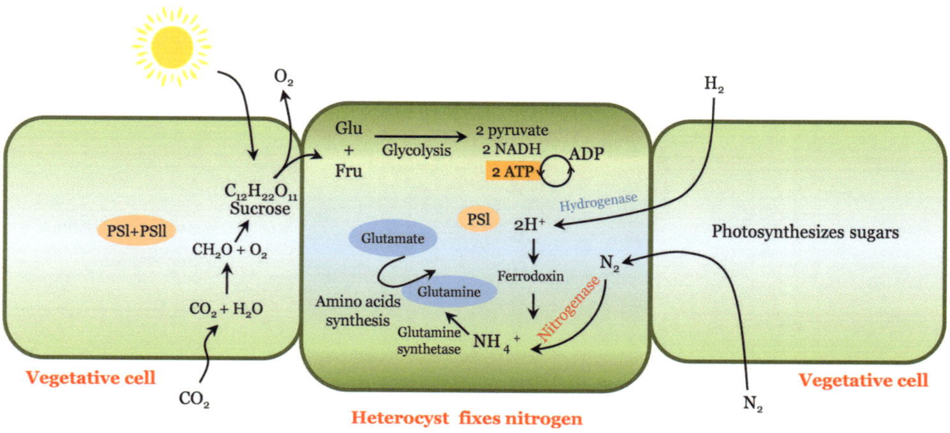

Figure 2.
Nitrogen fixation in cyanobacterial heterocyst cell.

cyanobacterium *Anabaena laxa* and the rhizobia *Mesorhizobium cicero* [18]. The assimilation of ammonium in cyanobacterial heterocysts is carried out by the enzyme glutamine synthetase. Heterocysts in the *Nostoc-Anthoceros* symbiosis showed a 3- to 4-fold reduction in Glutamine synthetase activity [19]. The reduction of abiotic stress and plant protection against diseases are factors that encourage the development of symbioses from a plant's side. Plants benefit from the general improvement of soil conditions.

4. The role of cyanobacteria in plant improvement

Cyanobacteria play an important role in improving plant growth and crop production. They are good bio-fertilizers, enhance solubilization and mobility of nutrients, and increase essential microelements in soil that are necessary for ion uptake, as well as stimulate plant growth due to their ability to produce bioactive compounds, such as phytohormones and other plant growth regulator substances, such as amino acids and polysaccharides (**Figure 3**).

4.1 Promoting plant growth

Cyanobacteria will actively promote seed germination, plant growth, and development due to their ability to produce some of the plant hormones, such as auxins, cytokinins, and gibberellins, by the genera *Anabaena, Anabaenopsis*, and *Calothrix* [20, 21]. Cyanobacteria have the ability to increase root and stem growth, dry weight, and yield in wheat [8, 20]. The cyanobacteria used in wheat cultivation showed effective results on the appearance of plants in terms of increasing plant height, dry weight, and a number of grains of the wheat crop, in addition to some important positive changes in increasing the bio-carbon content of the beneficial microbial mass [22]. The effects of cyanobacteria on rice crop growth have demonstrated that cyanobacterial inoculation can improve rice seed germination and growth parameters [23]. According to Osman et al. [24], the amount of growth-promoting secondary

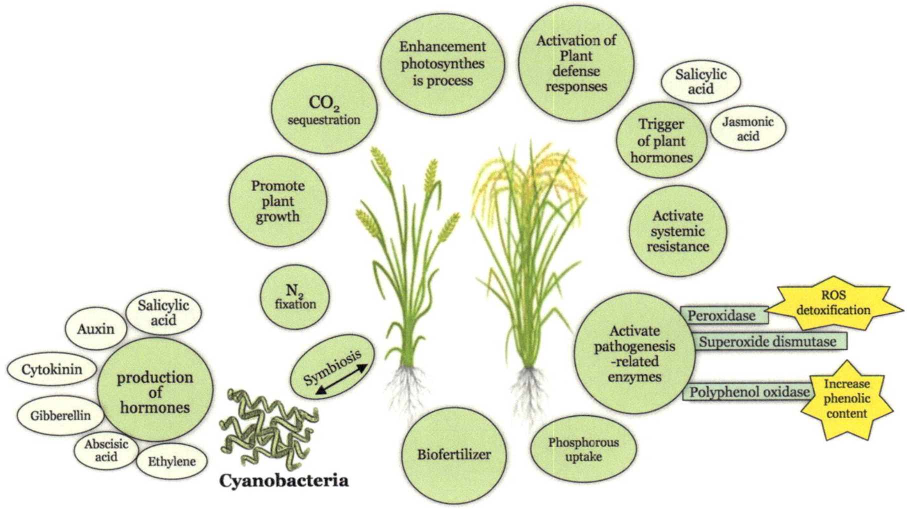

Figure 3.
The role of cyanobacteria in improving plant growth and stimulating the response of defense systems.

metabolites varies depending on the cyanobacterial strain. While *Oscillatoria angustissima* had higher quantities of gibberellic acid, and *Nostoc entophytum* had higher levels of auxin and cytokinin. Cyanobacterial extracts improved nutrient uptake, and plant development in lettuce, red beet [25], tomato [26], and cucumber [27]. In a broader sense, cyanobacteria are used as commercial bioinoculants to promote plant development because of their greater biodiversity, ability to survive in a variety of conditions, faster growth rate, and simpler nutritional requirements [28].

4.2 Nitrogen fixation

Nitrogen is the most important element needed for plant growth and is a key ingredient for successful cultivation on reclaimed land. Biological atmospheric nitrogen fixation by microorganisms is the main source of soil nitrogen [29]. Cyanobacteria have the ability to fix atmospheric nitrogen through specific cells called heterocysts that possess the nitrogenase enzyme. *Nostoc Linkia*, *Anabaena variabilis*, *Aulosira Fertilissima*, and *Calothrix* SP are the most efficient cyanobacteria in the soil's air nitrogen fixation [30]. Cyanobacteria get established permanently in the field after being applied for three to four subsequent crop seasons [31]. The growth characteristics of *Oryza sativa* were enhanced by the addition of *Nostoc commune* and *Nostoc carneum* as sources of cyanobacteria with chemical fertilizer [32]. Spraying the foliar of *Salix viminalis* L. three times with *Anabaena* sp. and *Microcystis aeruginosa* improved photosynthesis, stomatal conduction, and intracellular CO_2 concentration [33]. The application of *Nostoc entophytum* and *Oscillatoria angustissima* on *Pisum sativum* L. decreased the chemical fertilization to 50% [34]. The addition of cyanobacterium *Phormidium ambiguum* to sandy soil increased nitrate content by 15% more than the untreated soil after 90 days. Furthermore, the use of *Scytonema javanicum* improved the N content in slit loam, sandy loam, loamy sand, and sandy soils by 11, 10, 14, and 55%, respectively, the effect of cyanobacteria in biological crust formation and N supplementation for any sort of soil [35].

4.3 Bio fertility

In modern agriculture, microbes play a vital role in determining fertility and soil structure [36]. Cyanobacteria have potential use in agriculture as biofertilizers. Maintaining soil fertility using renewable bioresources is the main requirement of sustainable agriculture to reduce the need for synthetic fertilizers.

Among such resources, cyanobacteria are the most promising candidates. In the rhizosphere, cyanobacteria can be directly inoculated in the soil or can be used as a coating on seeds, but in both cases, their survival should be guaranteed. Although the use of agricultural chemical nitrogen fertilizers was a solution to all agricultural problems related to food production and increasing agricultural crop production, many environmental problems have arisen as a result of the excessive use of these chemical fertilizers in intensive farming systems. The high prices of chemical fertilizers have led to a decrease in the profit of agricultural crops, and the shortage of chemical fertilizers is a major problem facing farmers in developing countries, which makes researchers try to search for bio-alternatives to expensive chemical fertilizers [37]. Recently, there has been much interest in linking primary field crops in agriculture, especially cereal crops, such as wheat and rice, and organisms as a source of biofertilizers, such as cyanobacteria.

Due to the adaptation of cyanobacteria to most different environmental conditions, it is widely used in increasing soil fertility and improving soil quality and structure, so it has become one of the most important biofertilizers [38]. The effect

of cyanobacteria supplementation on growth, productivity, and physical properties of sandy soil under greenhouse conditions was tested. Sood et al. [39] found that there was a lot of ecological and metabolic diversity in cyanobacteria and that their structural-functional flexibility led to even more diversity. The use of cyanobacteria is one of the inexpensive applications in agriculture, which legalizes the use of chemical fertilizers. Cyanobacteria are one of the most important improvers that increase organic matter, amino acids, vitamins, and auxins in the soil, reduce soil salinity and phosphate deposits, and increase productivity in rice crops [40].

Cyanobacteria are emerging microorganisms for sustainable agricultural development. It can contribute about 20–30 kg of N per hectare, as well as soil organic matter, which is quite important for economically weak farmers who cannot invest in expensive chemical nitrogen fertilizers. The diazotroph group is the cyanobacteria most widely used for the development of biofertilizers and is capable of controlling the nitrogen deficiency in plants and improving the aeration of the soil and the water holding capacity. The most efficient nitrogen-fixing cyanobacteria are *Nostoc linkia*, *Anabaena variabilis*, *Aulosira fertilissima*, *Calothrix* sp., *Tolypothrix* sp., and *Scytonema* sp., which are normally present in the rice crop cultivation area.

4.4 Protection against abiotic stress

Abiotic stress on plants can be caused by a variety of factors, such as temperature, droughts, light, and soil-related factors, including salinity, presence of heavy metals, and soil acidity [41, 42]. Cyanobacteria induce diverse changes in response to elevated soil salinity by synthesis and accumulation of protective substances, maintaining low intracellular ion concentrations, and expression of so-called salt stress proteins [43]. *Anabaena torulosa* and *Anabaena* sp., exhibited anti-saline action by suppression of some expressed proteins, enhancement of other proteins, and expression of specialized salt stress proteins [44]. The effect of the extracellular products of *Scytonema hofmanni* on the growth of rice plants under salt stress was clearly demonstrated. These extracellular products made rice plants able to cope with stress caused by high salt concentrations. Comparison with the effects of plant gibberellic acid indicates that *S. hofmanni* produces gibberellin-like plant growth stimuli [45]. Another way to increase the sensitivity of plants to salinity stress is through the expression of cyanobacterial flavodoxin within them. This can induce multiple resistances in plants; it has been shown that it can reduce salt stress in *Medicago truncatula*. Adding cyanobacterium *Aphanothece* sp. and *Arthrospira maxima* led to improve tomato plant growth and increase the content of chlorophyll and nutrients essential content, such as nitrogen, phosphorous, and potassium, under saline stress [46]. Reduce the effects of salt stress on sweet pepper plants increase in growth, as well as in the water content of the plants by using a liquid extract of *Roholtiella* sp. [47].

The reduction of the harmful effect of abiotic stresses on plants was observed by cyanobacteria, which has a direct effect on the soil or an indirect effect through the activation of specific responses in plants [48]. Concerning salinity stress, the mechanisms of cyanobacteria depend on increasing the plant's ability to tolerate salinity through nitrogen fixation; the production of extracellular polysaccharides, compatible solutes, hormones, and antioxidative enzymes; the active export of ions; and the effects on the microbial community [49]. Rice plants showed an effective response to abiotic stress after treatment of rice roots with *Oscillatoria acuta* and *Plectonema boryanum*. That results in regular increases in the activity of peroxidase, phenylalanine ammonia-lyase, and phenylpropanoid [50]. Furthermore, rice plants

showed an increase in tolerance to salinity after inoculating roots with strains isolated from saline soils, such as *Nostoc calcicola*, *Nostoc linkia*, and *Anabaena variabilis* [51]. In salt-affected soils, *N. punctiforme* enhanced the physical composition, nutritional status, and microbial activity, leading to noticeably higher growth and yield [52].

Plant germination under drought stress can be enhanced by the use of cyanobacteria [53], moreover, it enhances the growth and development of plants in arid lands [54]. *Microcoleus* sp. and *Nostoc* sp. are capable of increasing germination and seedling growth of *Senna notabilis* and *Acacia hilliana* under drought stress [55]. Similar results were achieved in lettuce plants cultivated in barren soils after the addition of *Spirulina meneghiniana* and *Anabaena oryzae* [56].

Heavy metals can be effectively removed from agricultural soil and water by cyanobacteria [57]. Many cyanobacterial species, including *Anabaena variabilis*, *Nostoc muscorum*, *Aulosira fertilissimia*, and *Tolypothrix tenuis*, may absorb and remove Cr, Cu, Pb, and Zn [58], whereas *Oscillatoria* sp. and *Synechocystis* sp. can remove Cr, this was linked to increasing wheat growth [59]. Applying *Spirulina platensis* can hasten seed germination and boost plant growth by preventing Cd from moving from roots to shoots [60]. *Synechocystis* sp. and *Phormidium* sp. are capable of absorbing and removing systemic insecticide from the soil [61]. The addition of *S. platensis* in the soil can induce the biosynthesis of some amino acids, which can protect plants from the negative effects of the herbicide [62]. Cyanobacteria contribute to stimulating the release of plant hormones, such as salicylic acid or jasmonic acid, which have an effective role in protecting plants from biotic and abiotic stresses by stimulating gene expression of specific proteins [63]. Cyanobacteria lead to increased nitrogen and carbon content, state of soil aggregation, water retention, decrease in pH, exchangeable sodium, and decrease in heavy metals, as well as microbial flora reconstitution which in turn all have an effective role in reducing salt stress [49, 53].

5. The role of cyanobacteria in soil resilience

Soil health is seriously threatened in many parts of the world due to salinization, groundwater pollution from acidification, and excessive use of chemical fertilizers and pesticides. Cyanobacteria are essential for maintaining the health of the soil by enhancing soil physicochemical properties, including aggregation, aeration, and nutrient release patterns [20]. Additionally, cyanobacteria contribute to the fixation of nitrogen, excretion of biologically active compounds, increase soil biomass and organic matter, improve water-holding soil capacity, and improve soil phosphate bioavailability, moreover, cyanobacteria are alternative low-cost and eco-friendly that ensure soil sustainability (**Figure 4**).

5.1 Cyanobacteria improve physical properties of soil

In the upper crust of soil, the growth of cyanobacteria produces exopolysaccharides and extracellular polymers that alter the chemical composition and improve the physical properties of soil, which in turn promote beneficial microbial growth and strengthen soil structure [56]. Some cyanobacteria secrete mucilage or slime, which increases the availability of nutrients, and enhances soil structure that creates an ideal environment for the growth of advantageous microorganisms and plays a part in enhancing soil characteristics. The cyanobacteria *Nostoc muscorum* excrete exopolysaccharides and enhance saline soil stability [64]. Cyanobacteria can contribute to

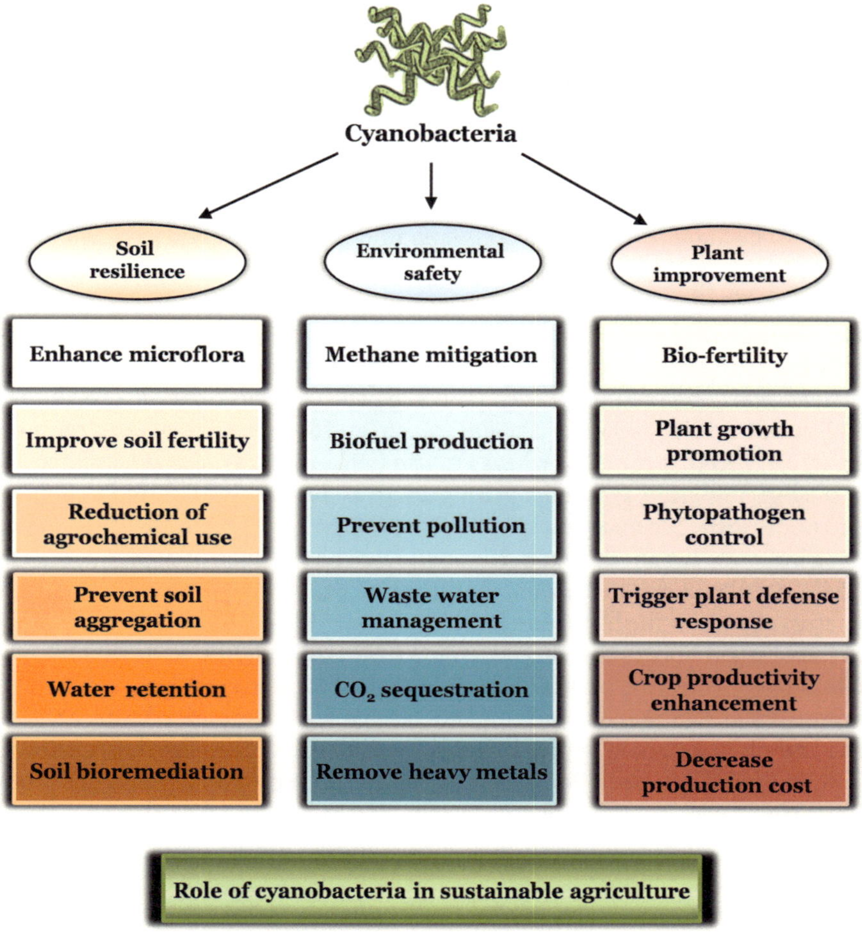

Figure 4.
An overview of the cyanobacterial role in sustainable agriculture and environmental safety.

the improvement and recovery of infertile soils by releasing holding and aggregation of soil particles together, the accumulation of organic content, and an increase in the water-holding capacity of the upper soil layer [65]. Rossi et al. [66] reported that the addition of cyanobacteria to the soil will improve soil properties and texture by adjusting soil stabilization, nutrients, moisture-holding capacity, and crust formation. The micromorphological characteristics of soil were improved after 6 weeks of the application of cyanobacteria combined with polysaccharides [67]. Chamizo et al. [35] demonstrated that the application of cyanobacteria can improve dry land functions through restoration and reestablishment. Cyanobacteria contribute to improving the properties of hard-to-cultivable lands, such as calcareous and saline soils, and making them suitable for cultivation.

5.2 Phosphorus uptake

Phosphorus is the second most important nutrient for plants after nitrogen. It is a crucial mineral for the growth and development of plants. It is one of the essential components of a live cell since it serves as the primary structural support for DNA,

RNA, and ATP [68]. Phosphate is frequently supplied to the soil in the form of phosphatic fertilizers. However, plants only use a small portion of this nutrient since a large portion of it is quickly converted to insoluble complexes in the soil that plants cannot utilize. With the help of phosphatase enzymes, cyanobacteria can solubilize and mobilize the insoluble organic phosphates present in the soil, for example, ferric phosphate, aluminum phosphate, tricalcium diphosphate, and hydroxyapatite into soluble forms and improve the bioavailability of phosphorus to the plants [69]. The use of cyanobacteria in crop fields plays a significant role in the mobilization of inorganic phosphates by extracellular phosphates and the excretion of organic acids. Cyanobacteria enhanced the decomposition and mineralization of phosphate and transformed it into readily available soluble organic phosphates.

5.3 Degradation of agrochemicals

Control of agricultural pests and weeds depends on the use of agrochemicals, for example, pesticides, fungicides, bactericides, insecticides, and herbicides. This leads to maintaining global food production by killing agricultural pests, but at the same time, these pesticides pollute the environment. Biological intervention for many beneficial microorganisms, including cyanobacteria, is involved in removing the chemical residues [70]. Cyanobacteria can be used to get rid of various pollutants, such as heavy metals, pesticides, chemical fertilizers, and crude oil [71]. Cyanobacteria are also able to remove heavy metals from water bodies and can reduce the increase in nitrates and phosphates from agricultural fields [72]. Intensive use of pesticides leads to an imbalance in the environmental system, especially in soil, water, and air. Currently, the use of beneficial microorganisms, especially cyanobacteria, is considered the best way to eliminate pesticides and chemicals that pollute agricultural soil. Cyanobacteria have the ability to break down pesticides at a faster rate. This requires some processes, such as adding the necessary nutrients or organic materials, to accelerate the rate of decreasing pollutants by the cyanobacteria, which have growth activities that exceed the chemical roads in addition to being environmentally friendly [73].

Among the different compounds used for agricultural applications the phosphorous-organic pesticide category. The random use of such chemicals causes many environmental problems. It also poses a great danger to other organisms, such as birds, fish, animals, and humans. As a result, it is highly recommended that these hazardous chemicals be removed from the environment in an appropriate manner. Cyanobacteria are one of the best applications of beneficial microorganisms because it breaks down toxic chemicals into nontoxic compounds. The widespread appearance of cyanobacteria in the polluted area is a contributing factor, making them a better candidate for biological decomposition [8].

6. The role of cyanobacteria in controlling phytopathogens

Plants can be attacked by bacteria, fungi, viruses, and nematodes at different stages of growth, causing severe harm to the root system, stem, leaves, and fruits. Chemical pesticides were the best approach to decrease the damage of these diseases, but pesticides have many negative effects over time. The use of biological alternatives, for example, cyanobacteria, has become a necessity to preserve the safety of the environment and the quality of crops. The major strategies used by cyanobacteria to

attack plant pathogens are antibiosis, the release of chemical compounds that may have the potential to inhibit a variety of phytopathogens, competition for space, and activation of plant defense responses. Cyanobacteria are distinguished by producing a huge number of bioactive substances (**Figure 5**). Thus, cyanobacteria provide a significant, safe alternative to avoid the harmful effects resulting from chemical control. It is a critical tool in sustainable agriculture [7].

Several plant fungi can be effectively controlled by cyanobacterial extracts, for example, *Fuarium* spp., *Verticillium* spp., *Alternaria* spp., *Penicillium* spp., *Botrytis cinerea*, *Rhizoctonia solani*, and *Sclerotinia sclerotiorum* [6]. Two orders of cyanobacteria, the Nostocales and Oscillatoriales, are very effective against fungal pathogens. Among Nostocales, two species, *Anabaena minutissima* and *Anabaena variabilis* are active in preventing the spread of airborne diseases [74, 75]. Airborne fungal pathogens produce a significant number of spores, which are considered the main source of spread. Therefore, inhibiting spore germination could play an effective role in controlling the disease and preventing secondary infection. Spraying of *A. minutissima* on cucurbit plants can reduce the symptoms of powdery mildew caused by *Podosphera xanthii*; also, infected areas of cucumber leaves and spore production decreased by 31% and 47%, respectively [75], while the disease was inhibited by 25% on zucchini [74]. *A. variabilis* has effective antibiosis against *R. solani* and *F. moniliforme* pathogens that infect tomato seedlings [76]. Also, *A. variabilis*, *N. linckia*, and *N. commune* have the same antibiosis effect on tomato wilt disease caused by *F. oxysporum* f. sp. *lycopersici* [76–78]. *Anabaena* sp. has an antibiosis against *P. xanthii*, which causes powdery mildew in zucchini plants [79]. *N. entophytum* and *N. muscorum* considerably decreased the activity of *R. solani* in soybean by an antibiosis mechanism [24]. Additionally, *Oscillatoria agardhii* has an antibiosis against *F. solani*, *Macrophomina phaseolina*, and *R. solani*, which cause the damping off disease in lupine seedlings [80].

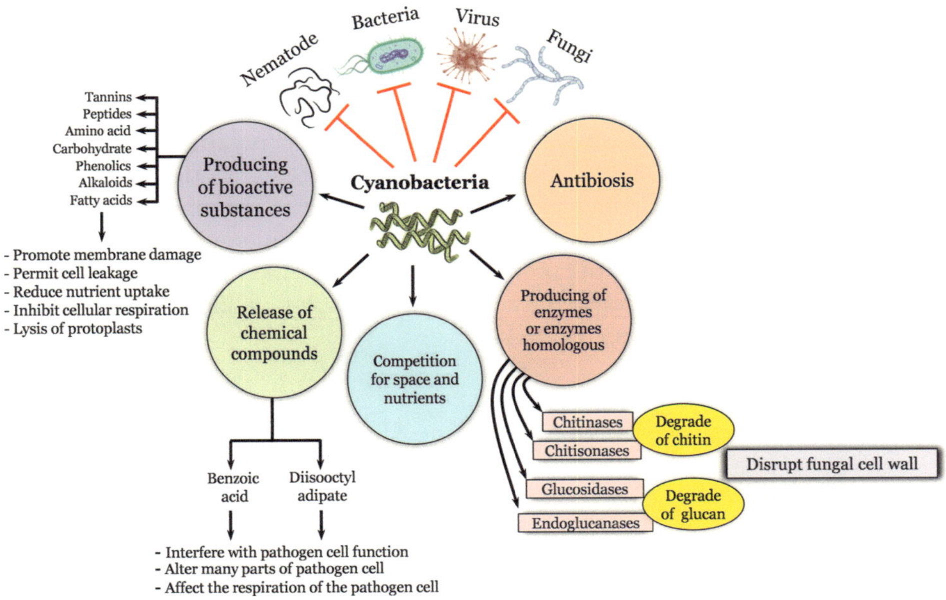

Figure 5.
Mechanisms of antimicrobial activity of cyanobacteria against phytopathogens.

The presence of diisooctyl adipate, extracted from *N. piscinale* and *A. variabilis*, is one strong indication that cyanobacteria contain chemical compounds active against *R. solani*, the causative agent of rice sheath blight, which causes severe damage in rice fields in China [81]. *Anabaena* spp., *Scytonema* spp., and *Nostoc* spp. have antifungal and toxic activity against soil-borne fungi [82]. *Rhizopus stolonifer*, *Phytophthora capsici*, *Pythium ultimum*, *Botrytis cinerea*, *Colletotrichum gloeosporoides*, *Fusarium oxysporum*, and *Alternaria solani* are all considerably inhibited by *Nostoc commune* methanolic extracts [83]. Additionally, methanolic extracts of *Spirulina platensis* effectively prevent the growth of *Helminthosporium* spp., *Alternaria brassicae*, *Aspergillus flavus*, and *Fusarium moniliforme* [84, 85].

Cyanobacteria can produce enzymes that directly act against the pathogen's cell wall. *Anabaena* sp. and *Calothrix elenkinii* can produce chitinases and chitisonases against pathogens, *F. moniliforme*, *F. solani*, *F. oxysporum*, *A. solani*, *M. phaseolina*, and *R. solani* and significantly reduce disease [86, 87]. Endoglucanases and glucosidases are two other enzymes that *Anabaena* sp. and *C. elenkinii* release. These enzymes can disrupt the cell walls of different plant pathogens by degradation of chitin and glucan, respectively [88]. In addition, Gupta et al. [89] reported that the antifungal properties of cyanobacteria are attributed to the production of endoglucanase, chitosanase homologs, and benzoic acid. Benzoic acid has the ability to interfere with fungal cell functioning, alter many parts of the cell, and has an effect on the respiration of the fungal cell [90]. Cyanobacteria can compete for space in the rhizosphere by forming biofilms at the roots and blocking sites of infection for soil pathogens, such as *Anabaena* sp., against *R. solani* in cotton roots [91].

On the other hand, they activate the defensive responses of the plant directly against fungal pathogens, such as *A. variabilis* or *A. laxa*, which enhance the activity of defense and pathogenesis-related enzymes in tomato roots against *F. oxysporum* f. sp. *lycopersici* [92], or by the activation of systemic resistance, such as *N. muscorum* and *A. oryzae,* that increase total phenol content and the activities of peroxidase, superoxide dismutase, and polyphenol oxidase enzymes in tomato leaves against *A. solani* [93].

The ability of cyanobacteria to combat various plant pathogenic bacteria and their ability to release compounds into the environment has been extensively studied [94]. The mechanism underlying the bactericidal action of cyanobacteria is attributed to the presence of tannins, amino acids, phenolics, alkaloids, carbohydrates, and fatty acids, which may cause bacterial membrane deterioration that eventually allows cells to leak, lowers nutrition intake, and prevents cellular respiration [95]. *Pseudomonas aeruginosa* is capable of infecting the roots of *A. thaliana* and *Ocimum basilicum*, causing plant death [96]. Nostoc sp. was effective in controlling *P. aeruginosa* due to the presence of long-chain fatty acids [97]. Additionally, *Anabaena flos-aquae* can completely suppress *Ralstonia solanacearum,* which causes brown rot disease in potatoes due to the production of antibiosis that is released into the environment [98]. Yanti et al. [99] found that cyanobacteria were able to stop *Ralstonia syzygii* subsp. *indonesiensis*, which is the cause of many vascular diseases in different crops.

Cyanobacteria possess antibiosis mechanisms against plant pathogenic nematodes that include paralysis, death, accelerating egg hatching, and inhibiting gall formation against plant harmful nematodes. *Heterodera cajani*, *Heterodera avenae*, *Meloidogyne graminicola*, *Meloidogyne incognita*, and *Rotylenchulus reniformis* can all be immobilized and killed by aqueous extracts of *Synechococcus nidulans* [100]. *Nostoc calcicola*, *Spirulina* sp., and *Anabaena oryzae* can lessen the quantity of nematode galls and egg masses in the cowpea rhizosphere [101]. *M. incognita* and *M. triticoryzae* are

nematostatically inhibited by *Aulosira fertilissima* [102]. Additionally, *M. incognita* eggs are inhibited from hatching by the cyanobacteria species *Anacystis nidulans*, *Oscillatoria fremyii*, and *Lyngbya* sp. [103]. Furthermore, *M. incognita* in the tomato rhizosphere can be eliminated by an aqueous extract of *Calothrix parietina* [104]. Additionally, *Microcoleus vaginatus* has the capacity to lower *M. incognita* populations in the tomato rhizosphere and reduce root galling [105]. By coming into touch with plant roots, cyanobacteria can trigger several nematode defense mechanisms in plants. In order to combat *M. incognita*, *S. platensis* increases the catalase activity in the roots of banana plants [106] and stimulates the production of the plant defense compound jasmonic acid in tomato plants [107].

7. Conclusion

The presence of cyanobacteria in the soil is a positive indicator of the availability of organic matter, the support of oxygen, and the synthesis of hormones, amino acids, and vitamins, in addition to increasing the solubility of phosphates and enhancing the efficiency of fertilizers in plants while reducing the soil content of oxidants and salinity. Additionally, it plays a crucial role in nitrogen fixation, which is a major source of plant nutrition. Moreover, cyanobacteria promote the production of plant hormones and play an important role in combating many phytopathogens as antifungal, antibacterial, antinematode, and antiviral. It is clear from the above that cyanobacteria play an integral role in the sustainability of agriculture, as they work to improve the physiology of the plant and protect it against abiotic stress and attack by phytopathogens. Therefore, cyanobacteria should be applied on a larger scale in modern agricultural systems.

Acknowledgements

The authors would like to thank the staff members of Plant Pathology Research Institute, Agricultural Research Center, Giza, Egypt.

Conflict of interest

The authors declare no conflict of interest.

All illustrations were designed and drawn by author Dr. Eman Elagamey.

Author details

Eman Elagamey[1*†], Magdi A.E. Abdellatef[1*†] and Hassan E. Flefel[2]

1 Plant Pathology Research Institute, Agricultural Research Center (ARC), Giza, Egypt

2 Department of Environmental Research, Theodor Bilharz Research Institute, Giza, Egypt

*Address all correspondence to: emanelagamey@arc.sci.eg, emanelagamey@yahoo.com, abdellatef72@arc.sci.eg and magdi_abdellatef@yahoo.com

† These authors contributed equally to this study.

References

[1] Pathak J, Maurya PK, Singh SP, Häder DP, Sinha RP. Cyanobacterial farming for environment friendly sustainable agriculture practices: Innovations and perspectives. Frontiers in Environmental Science. 2018;**6**:7. DOI: 10.3389/fenvs.2018.00007

[2] Dresselhaus T, Hückelhoven R. Biotic and abiotic stress responses in crop plants. Agronomy. 2018;**8**(11):267. DOI: 10.3390/agronomy8110267

[3] Van Oosten MJ, Pepe O, De Pascale S, Silletti S, Maggio A. The role of biostimulants and bioeffectors as alleviators of abiotic stress in crop plants. Chemical and Biological Technologies in Agriculture. 2017;**4**:5. DOI: 10.1186/s40538-017-0089-5

[4] Kumar VV. Biofertilizers and biopesticides in sustainable agriculture. In: Role of Rhizospheric Microbes in Soil. Singapore: Springer; 2018. pp. 377-398

[5] Enebe MC, Babalola OO. The impact of microbes in the orchestration of plants' resistance to biotic stress: A disease management approach. Applied Microbiology and Biotechnology. 2019;**103**:9-25. DOI: 10.1007/s00253-018-9433-3

[6] Righini H, Francioso O, Martel Quintana A, Roberti R. Cyanobacteria: A natural source for controlling agricultural plant diseases caused by fungi and oomycetes and improving plant growth. Horticulturae. 2022;**8**(1):58. DOI: 10.3390/horticulturae8010058

[7] Poveda J. Cyanobacteria in plant health: Biological strategy against abiotic and biotic stresses. Crop Protection. 2021;**141**:105450. DOI: 10.1016/j.cropro.2020.105450

[8] Chittora D, Mukesh M, Tansukh B, Prashant S. Cyanobacteria as a source of biofertilizers for sustainable agriculture. Biochemistry and Biophysics Reports. 2020;**22**:1-10. DOI: 10.1016/j.bbrep.2020.100737

[9] Vidal L, Ballot A, Azevedo SMFO, Padisák J, Welker M. Introduction to Cyanobacteria. In: Chorus I, Welker M, editors Toxic Cyanobacteria in Water. 2nd ed. London: World Health Organization; 2021. pp. 163-211. DOI: 10.1201/9781003081449

[10] Read N, Connell S, Adams DG. Nanoscale visualization of a fibrillar array in the cell wall of filamentous cyanobacteria and its implications for gliding motility. Journal of Bacteriology. 2007;**189**:7361-7366. DOI: 10.1128/JB.00706-07

[11] Berman-Frank I, Lundgren P, Falkowski P. Nitrogen fixation and photosynthetic oxygen evolution in cyanobacteria. Research in Microbiology. 2003;**154**:157-164. DOI: 10.1016/S0923-2508(03)00029-9

[12] Castenholz RW. General characteristics of the cyanobacteria. Bergey's Manual of Systematics of Archaea and Bacteria. 2015;**17**:1-23. DOI: 10.1002/9781118960608.cbm00019

[13] Adams DG, Duggan PS. Cyanobacteria–bryophyte symbioses. Journal of Experimental Botany. 2008;**59**:1047-1058. DOI: 10.1093/jxb/ern005

[14] Svircev Z, Tamas I, Nenin P, Drobac A. Co-cultivation of N2-fixing cyanobacteria and some agriculturally important plants in liquid and sand cultures. Applied Soil Ecology.

1997;**6**:301-308. DOI: 10.1016/S0929-1393(97)00022-X

[15] Múnera-Porras LM, García-Londoño S, Ríos-Osorio LA. Action mechanisms of plant growth promoting cyanobacteria in crops in situ: A systematic review of literature. International Journal of Agronomy. 2020;**2020**:2690410. DOI: 10.1155/2020/2690410

[16] Kollmen J, Strieth D. The beneficial effects of cyanobacterial co-culture on plant growth. Life. 2022;**12**(2):223. DOI: 10.3390/life12020223

[17] Gantar M, Kerby NW, Rowell P. Colonization of wheat (*Triticum vulgare* L.) by N2-fixing cyanobacteria: II. An ultrastructural study. The New Phytologist. 1991;**118**:485-492. DOI: 10.1111/j.1469-8137.1993.tb03842.x

[18] Prasanna R, Ramakrishnan B, Simranjit K, Ranjan K, Kanchan A, Hossain F, et al. Cyanobacterial and rhizobial inoculation modulates the plant physiological attributes and nodule microbial communities of chickpea. Archives of Microbiology. 2017;**199**:1311-1323. DOI: 10.1007/s00203-017-1405-y

[19] Joseph CM, Meeks JC. Regulation of expression of glutamine synthetase in a symbiotic nostoc strain associated with *Anthoceros punctatus*. Journal of Bacteriology. 1987;**169**:2471-2475

[20] Singh JS, Kumar A, Rai AN, Singh DP. Cyanobacteria: A precious bio-resource in agriculture, ecosystem, and environmental sustainability. Frontiers in Microbiology. 2016;**7**:529. DOI: 10.3389/fmicb.2020.548410

[21] Joshi H, Shourie A, Singh A. Cyanobacteria as a source of biofertilizers for sustainable agriculture. In: Advances in Cyanobacterial Biology. Academic Press; 2020. pp. 385-396. DOI: 10.1016/B978-0-12-819311-2.00025-5

[22] Karthikeyan N, Radha P, Nb L, Kaushik BD. Evaluating the potential of plant growth promoting cyanobacteria as inoculants for wheat. European Journal of Soil Biology. 2007;**43**:23-30. DOI: 10.1016/j.ejsobi.2006.11.001

[23] Yk J, Panpatte DG, Vyas RV. In: Panpatte D, Jhala Y, Vyas R, Shelat H, editors. Cyanobacteria: Source of Organic Fertilizers for Plant Growth, Microorganisms for Green Revolution. Microorganisms for Sustainability. Singapore: Springer; 2017. p. 6. DOI: 10.1007/978-981-10-6241-4_13

[24] Osman MEH, El-Sheekh MM, Metwally AM, Ismail AA, Ismail MM. Antagonistic activity of some fungi and cyanobacteria species against *Rhizoctonia solani*. International Journal of Plant Pathology. 2011;**2**(3):101-114. DOI: 10.3923/ijpp.2011.101.114

[25] Mógor ÁF, Ördög V, Pereira Lima GP, Molnár Z, Mógor G. Biostimulant properties of cyanobacterial hydrolysate related to polyamines. Journal of Applied Phycology. 2018;**30**:453-460. DOI: 10.1007/s10811-017-1242-z

[26] Supraja KV, Behera B, Balasubramanian P. Efficacy of microalgal extracts as biostimulants through seed treatment and foliar spray for tomato cultivation. Industrial Crops and Products. 2020;**151**:112453. DOI: 10.1016/j.indcrop.2020.112453

[27] Toribio AJ, Suárez-Estrella F, Jurado MM, López MJ, López-González JA, Moreno J. Prospection of cyanobacteria producing bioactive substances and their application as potential phytostimulating agents. Biotechnology Reports. 2020;**26**:e00449. DOI: 10.1016/j.btre.2020.e00449

[28] Ruffing AM. Engineered cyanobacteria: Teaching an old bug new tricks. Bioengineered Bugs. 2011;**2**:136-149. DOI: 10.4161/bbug.2.3.15285

[29] Mahanty T, Bhattacharjee S, Goswami M, Bhattacharyya P, Das B, Ghosh A, et al. Biofertilizers: A potential approach for sustainable agriculture development. Environmental Science and Pollution Research. 2017;**24**(4):3315-3335. DOI: 10.1007/s11356-016-8104-0

[30] Prasad RC, Prasad BN. Cyanobacteria as a source biofertilizer for sustainable agriculture in Nepal. Journal in Plant Science Botanica Orientalis. 2001;**1**:127-133

[31] Esteves-Ferreira AA, Cavalcanti JHF, Vaz MGMV, Alvarenga LV, Nunes-Nesi A, Araújo WL. Cyanobacterial nitrogenases: Phylogenetic diversity, regulation and functional predictions. Genetics and Molecular Biology. 2017;**40**(1):261-275. DOI: 10.1590/1678-4685-GMB-2016-0050

[32] Chittapun S, Limbipichai S, AmnuaysinN, BoonkerdR, CharoensookM. Effects of using cyanobacteria and fertilizer on growth and yield of rice, Pathum Thani I: A pot experiment. Journal of Applied Phycology. 2018;**30**(1):79-85. DOI: 10.1007/s10811-017-1138-y

[33] Grzesik M, Romanowska-Duda Z, Kalaji HM. Effectiveness of cyanobacteria and green algae in enhancing the photosynthetic performance and growth of willow (*Salix viminalis* L.) plants under limited synthetic fertilizers application. Photosynthetica. 2017;**55**(3):510-521. DOI: 10.1007/s11099-017-0716-1

[34] Osman MEH, El-Sheekh MM, El-Naggar AH, Gheda SF. Effect of two species of cyanobacteria as biofertilizers on some metabolic activities, growth, and yield of pea plant. Biology and Fertility of Soils. 2010;**46**(8):861-875. DOI: 10.1007/s00374-010-0491-7

[35] Chamizo S, Mugnai G, Rossi F, Certini G, De Philippis R. Cyanobacteria inoculation improves soil stability and fertility on different textured soils: Gaining insights for applicability in soil restoration. Frontiers of Environmental Science. 2018;**6**(49):00049. DOI: 10.3389/fenvs.2018

[36] Singh S. A review on possible elicitor molecules of cyanobacteria: Their role in improving plant growth and providing tolerance against biotic or abiotic stress. Journal of Applied Microbiology. 2014;**117**:1221-1244. DOI: 10.1111/jam.12612

[37] Abdelfattah MS, Elsadany AY, Doha N. The use of cyanobacteria as biofertilizer in wheat cultivation under different nitrogen rates. Natural Science. 2018;**16**:30-35. DOI: 10.7537/marsnsj160418.06

[38] Dhar DW, Prasanna R, Singh BV. Comparative performance of three carrier based blue green algal biofertilizers for sustainable rice cultivation. Journal of Sustainable Agriculture. 2007;**30**:41-50. DOI: 10.1300/j064v30n02_06

[39] Sood A, Renuka N, Prasanna R, Ahluwalia AS. Cyanobacteria as potential options for wastewater treatment. In: Phytoremediation. Cham: Springer; 2015. pp. 83-93. DOI: 10.1007/978-3-319-10969-5_8

[40] Kaushik BD. Use of blue-green algae and azolla biofertilizers in rice cultivation and their influence on soil properties. In: Jain PC, editor. Microbiology and Biotechnology for sustainable development. New Delhi, India: CBS Publishers & Distributors; 2004. pp. 166-184

[41] Cramer GR, Urano K, Delrot S, Pezzotti M, Shinozaki K. Effects of abiotic stress on plants: A systems biology

perspective. BMC Plant Biology. 2011;**11**: 163. DOI: 10.1186/1471-2229-11-163

[42] Gr S, Yadav RK, Chatrath A, Gerard M, Tripathi K, Govindsamy V, et al. Perspectives on the potential application of cyanobacteria in the alleviation of drought and salinity stress in crop plants. Journal of Applied Phycology. 2021;**33**:3761-3778. DOI: 10.1007/s10811-021-02570-5

[43] Singh NK, Dhar DW. Cyanobacterial reclamation of salt-affected soil. In: Lichtfouse E, editor. Genetic Engineering, Biofertilisation, Soil Quality and Organic Farming. Dordrecht, The Netherlands: Springer; 2010. pp. 243-275

[44] Apte SK, Bhagwat AA. Salinity-stress-induced proteins in two nitrogen-fixing anabaena strains differentially tolerant to salt. Journal of Bacteriology. 1989;**171**(2):909-915. DOI: 10.1128/jb.171.2.909-915.1989

[45] Rodríguez AA, Stella AM, Storni MM, Zulpa G, Zaccaro MC. Effects of cyanobacterial extracellular products and gibberellic acid on salinity tolerance in *Oryza sativa* L. Saline Systems. 2006;**2**:7. DOI: 10.1186/1746-1448-2-7

[46] Mutale-Joan C, Rachidi F, Mohamed HA, Mernissi NE, Aasfar A, Barakate M, et al. Microalgae-cyanobacteria-based biostimulant effect on salinity tolerance mechanisms, nutrient uptake, and tomato plant growth under salt stress. Journal of Applied Phycology. 2021;**33**:3779-3795. DOI: 10.1007/s10811-021-02559-0

[47] Coba de la Peña T, Redondo FJ, Manrique E, Lucas MM, Pueyo JJ. Nitrogen fixation persists under conditions of salt stress in transgenic *Medicago truncatula* plants expressing a cyanobacterial flavodoxin. Plant Biotechnology Journal. 2010;**8**(9):954-965. DOI: 10.1111/j.1467-7652.2010.00519.x

[48] Singh JS. Cyanobacteria: A vital bio-agent in eco-restoration of degraded lands and sustainable agriculture. Climate Change and Environmental Sustainability. 2014;**2**:133-137

[49] Li H, Zhao Q, Huang H. Current states and challenges of salt-affected soil remediation by cyanobacteria. Science of the Total Environment. 2019;**669**:258-272. DOI: 10.1016/j.scitotenv.2019.03.104

[50] Singh DP, Prabha R, Yandigeri MS, Arora DK. Cyanobacteria-mediated phenylpropanoids andphytohormones in rice (*Oryza sativa*) enhance plant growth and stress tolerance. Antonie Van Leeuwenhoek. 2011;**100**(4):557-568. DOI: 10.1007/s10482-011-9611-0

[51] El Sheekh MM, Zayed MA, Elmossel FK, Hassan RS. Effect of cyanobacteria isolates on rice seeds germination in saline soil. Baghdad Science Journal. 2018;**15**(1):16-21. DOI: 10.21123/bsj.2018.15.1.0016

[52] Nisha R, Kiran B, Kaushik A, Kaushik CP. Bioremediation of salt affected soils using cyanobacteria in terms of physical structure, nutrient status and microbial activity. International Journal of Environmental Science and Technology. 2018;**15**:571-580. DOI: 10.1007/s13762-017- 1419-7

[53] Rai LC, Singh S, Pradhan S. Biotechnological potential of naturally occurring and laboratory-grown microcystis in biosorption of Ni^{2+} and Cd^{2+}. Current Science. 1998;**74**:461-464

[54] Chua M, Erickson TE, Merritt DJ, Chilton AM, Ooi MKJ, Muñoz-Rojas M. Bio-priming seeds with cyanobacteria: Effects on native plant growth and soil properties. Restoration Ecology. 2020;**28**:S168-S176. DOI: 10.1111/rec.13040

[55] Muñoz-Rojas M, Chilton A, Liyanage GS, Erickson TE, Merritt DJ, Neilan BA, et al. Effects of indigenous soil cyanobacteria on seed germination and seedling growth of arid species used in restoration. Plant and Soil. 2018;**429**:91-100. DOI: 10.1007/s11104-018-3607-8

[56] Ibraheem IBM. Cyanobacteria as alternative biological conditioners for bioremediation of barren soil. Egyptian Journal of Phycology. 2007;**8**:99-107. DOI: 10.21608/egyjs.2007.114548

[57] Priyanka KC, Chatterjee A, Wenjing W, Yadav D, Singh PK. Cyanobacteria: Potential and role for environmental remediation. In: Abatement of Environmental Pollutants. Elsevier; 2020. pp. 193-202. DOI: 10.1016/B978-0-12-818095-2.00010-2

[58] Kaur R, Goyal D. Heavy metal accumulation from coal fly ash by cyanobacterial biofertilizers. Particulate Science and Technology. 2018;**36**:513-516. DOI: 10.1080/02726351.2017.1398794

[59] Faisal M, Hameed A, Hasnain S. Chromium-resistant bacteria and cyanobacteria: Impact on Cr (VI) reduction potential and plant growth. Journal of Industrial Microbiology & Biotechnology. 2005;**32**:615-621. DOI: 10.1007/s10295-005-0241-2

[60] Seifikalhor M, Hassani SB, Aliniaeifard S. Seed priming by cyanobacteria (*Spirulina platensis*) and salep gum enhances tolerance of maize plant against cadmium toxicity. Journal of Plant Growth Regulation. 2020;**39**(3):1009-1021. DOI: 10.1007/s00344-019-10038-7

[61] Aminfarzaneh H, Duygu E. The effect of salicylic acid and triacontanol on biomass production and imidaclopirid removal capacity by cyanobacteria. Communications Faculty of Sciences University of Ankara Series C. 2010;**22**:15-31. DOI: 10.1501/Commuc_0000000174

[62] Osman MEH, Abo-Shady AM, El-Nagar MM. Cyanobacterial *Arthrospira* (*Spirulina platensis*) as safener against harmful effects of fusilade herbicide on faba bean plant. Rendiconti Lincei. 2016;**27**:455-462. DOI: 10.1007/s12210-015-0498-y

[63] Khan MR, Syeed S, Nazar R, Anjum NA. An insight into the role of salicylic acid and jasmonic acid in salt stress tolerance. In: Khan NA, Nazar R, Iqbal N, Anjum NA, editors. Phytohormones and Abiotic Stress Tolerance in Plants. Berlin/Heidelberg, Germany: Springer; 2012. pp. 277-300. DOI: 10.1007/978-3-642-25829-9_12

[64] Caire G, de Cano SM, de Mulé ZMC, Palma RM, Colombo K. Exopolysaccharides of *Nostoc muscorum* ag. (cyanobacteria) in the aggregation of soil particles. Journal of Applied Phycology. 1997;**9**:249-253. DOI: 10.1023/A:1007994425799

[65] Garlapati D, Chandrasekaran M, Devanesan A, Mathimani T, Pugazhendhi A. Role of cyanobacteria in agricultural and industrial sectors: An outlook on economically important byproducts. Applied Microbiology and Biotechnology. 2019;**103**:4709-4721. DOI: 10.1007/s00253-019-09811-1

[66] Rossi F, Li H, Liu Y, De Philippis R. Cyanobacterial inoculation (cyanobacterisation): Perspectives for the development of a standardized multifunctional technology for soil fertilization and desertification reversal. Earth Science Reviews. 2017;**171**:28-43. DOI: 10.1016/j.earscirev.2017.05.006

[67] Malam Issa O, Défarge C, BissonnaisY L, Marin B, Duval O,

Bruand A, et al. Effects of the inoculation of cyanobacteria on the microstructure and the structural stability of a tropical soil. Plant and Soil. 2007;**290**(1):209-219. DOI: 10.1007/s11104-006-9153-9

[68] Cordell D, Rosemarin A, Schröder JJ, Smit AL. Towards global phosphorus security: A systems framework for phosphorus recovery and reuse options. Chemosphere. 2011;**84**:747-758. DOI: 10.1016/j.chemosphere.2011.02.032

[69] Rai AN, Singh AK, Syiem MB. Plant growth-promoting abilities in cyanobacteria. In: Cyanobacteria. Amsterdam, The Netherlands: Elsevier; 2019. pp. 459-476. DOI: 10.1016/B978-0-12-814667-5.00023-4

[70] Subashchandrabose SR, Balasubramanian R, Mallavarapu M, Kadiyala V, Ravi N. Mixotrophic cyanobacteria and microalgae as distinctive biological agents for organic pollutant degradation. Environment International. 2013;**51**:59-72. DOI: 10.1016/j.envint.2012.10.007

[71] Singh R, Parul P, Madhulika S, Andrzej B, Jitendra K, Samiksha S, et al. Uncovering potential applications of cyanobacteria and algal metabolites in biology, agriculture and medicine: Current status and future prospects. Frontiers in Microbiology. 2017;**8**:1-37. DOI: 10.3389/fmicb.2017.00515

[72] Kumar D, Pandey LK, Gaur JP. Metal sorption by algal biomass: From batch to continuous system. Algal Research. 2016;**18**:95-109. DOI: 10.1016/j.algal.2016.05.026

[73] Verma JP, Jaiswal DK, Sagar R. Pesticide relevance and their microbial degradation: A-state-of-art. Reviews in Environmental Science and Biotechnology. 2014;**13**:429-466. DOI: 10.1007/s11157-014-9341-7

[74] Roberti R, Galletti S, Burzi PL, Righini H, Cetrullo S, Perez C. Induction of defence responses in zucchini (*Cucurbita pepo*) by *anabaena* sp. water extract. Biological Control. 2015;**82**:61-68. DOI: 10.1016/j.biocontrol.2014.12.006

[75] Righini H, Somma A, Cetrullo S, D'Adamo S, Flamigni F, Martel Quintana A, et al. Inhibitory activity of aqueous extracts from *Anabaena minutissima*, *Ecklonia maxima* and *Jania adhaerens* on the cucumber powdery mildew pathogen in vitro and in vivo. Journal of Applied Phycology. 2020;**32**:3363-3375. DOI: 10.1007/s10811-020-02160-x

[76] Chaudhary V, Prasanna R, Nain L, Dubey SC, Gupta V, SinghR J, et al. Bioefficacy of novel cyanobacteriaamended formulations in suppressing damping off disease in tomato seedlings. World Journal of Microbiology and Biotechnology. 2012;**28**(12):3301-3310. DOI: 10.1007/s11274-012-1141-z

[77] Kim J, Kim JD. Inhibitory effect of algal extracts on mycelia growth of the tomato-wilt pathogen, *Fusarium oxysporum* f. sp. *lycopersici*. Mycobiology. 2008;**36**(4):242-248. DOI: 10.4489/MYCO.2008.36.4.242

[78] Alwathnani HA, Perveen K. Biological control of fusarium wilt of tomato by antagonist fungi and cyanobacteria. African Journal of Biotechnology. 2012;**11**:1100-1105. DOI: 10.5897/AJB11.3361

[79] Roberti R, Righini H, Reyes CP. Activity of seaweed and cyanobacteria water extracts against *Podosphaera xanthii* on zucchini. Italian Journal of Mycology. 2016;**45**:66-77. DOI: 10.6092/issn.2531-7342/6425

[80] Abdel-Monaim MF, Mazen MM, Atwa MAAM. Effect of cyanobacteria

on reducing damping-off and root rot incidence in lupine plants, New Valley Governorate, Egypt. Microbiology Research Journal International. 2016;**16**:1-14. DOI: 10.9734/BMRJ/2016/27883

[81] Zhou Y, Bao J, Zhang D, Li Y, Li H, He H. Effect of heterocystous nitrogen-fixing cyanobacteria against rice sheath blight and the underlying mechanism. Applied Soil Ecology. 2020;**153**:103580. DOI: 10.1016/j.apsoil.2020.103580

[82] Bloor S, England RR. Antibiotic production by the cyanobacterium *Nostoc muscorum*. Journal of Applied Phycology. 1989;**1**(4):367-372

[83] Kim JD. Screening of cyanobacteria (blue-green algae) from rice paddy soil for antifungal activity against plant pathogenic fungi. Microbiology. 2006;**34**:138-142. DOI: 10.4489/MYCO.2006.34.3.138

[84] Abedin RMA, Taha HM. Antibacterial and antifungal activity of cyanobacteria and green microalgae. Evaluation of medium components by Plackett-Burman design for antimicrobial activity of *Spirulina platensis*. Global Journal of Biotechnology & Biochemistry. 2008;**3**(1):22-31

[85] Karthika N, Muruganandam A. Bioactive compounds and antimicrobial activity of cyanobacteria from south east coast of India. International Journal of Current Research in Life Sciences. 2019;**8**(1):3027-3030

[86] Prasanna R, Saxena G, Singh B, Ranjan K, Buddhadeo R, Velmourougane K, et al. Mode of application influences the biofertilizing efficacy of cyanobacterial biofilm formulations in chrysanthemum varieties under protected cultivation. Open Agriculture. 2018;**3**:478-489. DOI: 10.1515/opag-2018-0053

[87] Mahawar H, Prasanna R, Gogoi R. Elucidating the disease alleviating potential of cyanobacteria, copper nanoparticles and their interactions in *Fusarium solani* challenged tomato plants. Plant Physiology Reports. 2019;**24**:533-540. DOI: 10.1007/s40502-019-00490-8

[88] Elagamey E, Abdellatef MAE, Arafat MY. Proteomic insights of chitosan mediated inhibition of *fusarium oxysporum* f. sp. *cucumerinum*. Journal of Proteomics. 2022;**260**:104560. DOI: 10.1016/j.jprot.2022.104560

[89] Gupta V, Natarajan C, Kumar K, Prasanna R. Identification and characterization of endoglucanases for fungicidal activity in *Anabaena laxa* (cyanobacteria). Journal of Applied Phycology. 2011;**23**(1):73-81. DOI: 10.1007/s10811-010-9539-1

[90] Mohamed SA, Kordali Ş, Korkmaz M. Evaluation of the effect of benzoic acid on some plant pathogenic fungi. International Journal of Agricultural and Natural Sciences. 2018;**1**(1):3-5

[91] Prasanna R, Babu S, Bidyarani N, Kumar A, Triveni S, Monga D, et al. Prospecting cyanobacteria-fortified composts as plant growth promoting and biocontrol agents in cotton. Experimental Agriculture. 2014;**51**:42-65. DOI: 10.1017/S0014479714000143

[92] Prasanna R, Chaudhary V, Gupta V, Babu S, Kumar A, Singh R, et al. Cyanobacteria mediated plant growth promotion and bioprotection against fusarium wilt in tomato. European Journal of Plant Pathology. 2013;**36**:337-353. DOI: 10.100 7/s10658-013-0167-x

[93] Attia MS, Sharaf AEMA, Zayed AS. Protective action of some bio-pesticides against early blight disease caused by *Alternaria solani* in tomato plant.

IJISET. 2017;**4**:67-94. DOI: 10.13140/RG.2.2.12528.56322

[94] Shao J, Peng L, Luo S, Yu G, Gu JD, Lin S, et al. First report on the allelopathic effect of *Tychonema bourrellyi* (cyanobacteria) against *Microcystis aeruginosa* (cyanobacteria). Journal of Applied Phycology. 2013;**25**:1567-1573. DOI: 10.1007/s10811-012-9969-z

[95] Smith VJ, Desbois AP, Dyrynda EA. Conventional and unconventional antimicrobials from fish, marine invertebrates and micro-algae. Marine Drugs. 2010;**8**(4):1213-1262. DOI: 10.3390/md8041213

[96] Walker TS, Bais HP, Déziel E, Schweizer HP, Rahme LG, Fall R, et al. *Pseudomonas aeruginosa*-plant root interactions. Pathogenicity, biofilm formation, and root exudation. Plant Physiology. 2004;**134**:320-331. DOI: 10.1104/pp.103.027888

[97] Catarina Guedes A, Barbosa CR, Amaro HM, Pereira CI, Xavier MF. Microalgal and cyanobacterial cell extracts for use as natural antibacterial additives against food pathogens. International Journal of Food Science and Technology. 2011;**46**:862-870. DOI: 10.1111/j.1365-2621.2011.02567.x

[98] Mikhail MS, Hussein BA, Messiha NA, Morsy KMM, Youssef MM. Using of cyanobacteria in controlling potato brown rot disease. International Journal of Engineering Research. 2016;**7**:274-281

[99] Yanti Y, Hamid H, Syarif Z. Screening of indigenous rhizospheric cyanobacteria as potential growth promotor and biocontrol of *Ralstonia syzygii* subsp. *indonesiensis* on chili. International Journal of Agriculture Environment and Biotechnology. 2019;**4**:1665-1672. DOI: 10.22161/ijeab.46.5

[100] Holajjer P, Kamra A, Gaur HS, Dhar DW. *In vitro* nematicidal activity of a terrestrial cyanobacterium, *Synechococcus nidulans*, towards plant-parasitic nematodes. Nemato. 2012;**14**:85-92. DOI: 10.1163/138855411X578879

[101] Youssef MMA, Ali MS. Management of *Meloidogyne incognita* infecting cowpea by using some native blue green algae. Anzeiger für Schädlingskunde, Pflanzenschutz, Umweltschutz. 1998;**71**:15-16. DOI: 10.1007/BF02770565

[102] Chandel ST. Nematicidal activity of the cyanobacterium, *Aulosira fertilissima* on the hatch of *Meloidogyne triticoryzae* and *Meloidogyne incognita*. Archives of Phytopathology and Plant Protection. 2009;**42**:32-38. DOI: 10.1080/03235400600914363

[103] Sharma HK, Gaur HS. Hatch inhibition of *Meloidogyne incognita* by aqueous extracts and exudates of five species of cyanobacteria. Nematologia Mediterranea. 2008;**36**:99-102

[104] Hashem M, Abo-Elyousr KA. Management of the root-knot nematode *Meloidogyne incognita* on tomato with combinations of different biocontrol organisms. Crop Protection. 2011;**30**:285-292. DOI: 10.1016/j.cropro.2010.12.009

[105] Khan Z, Park SD, Shin SY, Bae SG, Yeon IK, Seo YJ. Management of *Meloidogyne incognita* on tomato by root-dip treatment in culture filtrate of the blue-green alga. *Microcoleus vaginatus*. Bioresource Technology. 2005;**96**:1338-1341. DOI: 10.1016/j.biortech.2004.11.012

[106] Hamouda R, Al-Saman M, El-Ansary M. Effect of *Saccharomyces cerevisiae* and *Spirulina platensis* on suppressing root-knot nematode, *Meloidogyne incognita* infecting banana plants under greenhouse conditions.

Egyptian Journal of Agronematology. 2019;**18**:90-102. DOI: 10.21608/ejaj.2019.52593

[107] Sharaf AEMA, Kailla AM, Attia MS, Nofal MM. Induced resistance in tomato plants against root knot nematode using biotic and abiotic inducers. IJARBS. 2016;**3**:31-46. DOI: 10.22192/ijarbs.2016.03.11.004

Chapter 5

Cyanobacteria as a Source of Biodegradable Plastics

Mohanasundaram Yogeswar, Natarajan Valle and Arumugam Nagarajan

Abstract

Polyhydroxyalkanoates (PHAs) are a group of biopolymers produced from various microorganisms that attracted many researchers for their use as a substitute for conventional petrochemical plastics. PHA possesses similar material properties to petrochemical plastics with the added benefits of biocompatibility, biodegradability, hydrophobicity, thermoplasticity, piezoelectricity, and stereospecificity. The first discovery of PHA production in cyanobacteria was in 1969, and the commercialization of PHA produced from cyanobacteria is not feasible to date. The difficulty with the commercial production of cyanobacterial PHA is due to the low biomass production and lower PHA accumulation than the heterotrophic bacteria. The biosynthesis of PHA, production of cyanobacterial PHA, and strategies to improve the production of PHA and commercialization are discussed in this chapter.

Keywords: cyanobacteria, Polyhydroxyalkanoates, biodegradable polymers, bioplastics, bioprocess, PHB, P(HB-HV), PHA properties

1. Introduction

Bioplastics are a type of plastic that can be produced from natural materials like plant starches and oils. By 2025, it is anticipated that the amount of petroleum used to produce plastic would have decreased by 15–20% due to the use of bioplastics, which are made from plants. Asia and Europe will hold the biggest market share for bioplastics by 2025. Asia will make up 32% of the market, followed by Europe at 31% and the United States at 28%. The market for bioplastics is now growing at a rate of 10% per year, accounting for 10–15% of the entire plastics business in 2016 and increased to 25–30% in 2020 [1]. Synechocystis, Spirulina, Anabaena, and Nostoc muscorum are cyanobacteria that can serve as bio-factories for the production of biofuel and bioplastic. They can produce biopolymers like polyhydroxybutyrate (PHB) and polyhydroxyalkanoates (PHAs), among other copolymers, that are both affordable and sustainable [2].

Recent bioplastics like Bio-PET are only called biobased since their monomers are made from corn, but the polymerization process is chemical, and the final polymer has the same qualities as traditional PET, making it nondegradable [3]. Scytonema geitleri and other cyanobacterial species can store internal poly-hydroxybutyrate granules for

energy and carbon reserve when under stress. The environmentally benign and biodegradable PHB can then be collected and utilized to create biocompatible thermoplastics [4]. Polyhydroxyalkanoates (PHAs) are a type of polymer produced by cyanobacteria. PHAs are lipid compounds that a variety of microbes accumulate when there are abundant carbon sources present. They can be used for a variety of purposes, including the creation of bioplastics [5]. Cyanobacteria need only a small amount of nutrients to develop, and they produce PHAs through oxygenated photosynthesis [6].

Biochemical processes can naturally recycle bioplastics manufactured from renewable resources, reducing the need for fossil fuels and preserving the environment. Bioplastics are therefore environmentally friendly, generally biodegradable, and biocompatible. In many industrial applications today, including horticulture, food packaging, hygiene, AND composting bags, bioplastics have become essential. Additionally, bioplastics are utilized in biological, structural, electrical, and other consumer goods. With the demand for plastic usage increasing globally, a lot of research is being done to investigate green materials and novel processing techniques.

Chlorosis is the term for the dormant state that occurs when nutrients are scarce, such as nitrogen. During chlorosis, cyanobacteria deteriorate their photosynthetic machinery. Beyond this breakdown, there is a significant buildup of glycogen for the storage of carbon and energy. The process ends with the cells starting to break down the glycogen and turn it into PHB [7]. The sole PHA synthesized under the described photoautotrophic state out of the several PHAs is PHB. It is possible to add organic carbon precursors like valerate to make the additional short-chain-length PHAs (scl-PHAs), such as P(3HB-co-3 HV). Long-chain-length PHA or mcl-PHA have not yet been found in cyanobacteria. The most effective known catalyst for PHB synthesis in cyanobacteria is nitrogen restriction [8]. It has been noted that elements including culture conditions, such as N, P, light exposure, and CO2 dynamics, have an impact on cyanobacteria's ability to produce PHA. Additionally, it has been noted that additional elements including two-stage (growth and PHA accumulation) processes, metabolic inhibitors for other pathways, and bioengineering have a favorable effect on PHA production [9–13].

1.1 Types of bioplastics

Following is a classification of bioplastics based on the wide definition:

i. Starch-Based Bioplastics—Starch-based polymers are defined as those that contain either natural or modified starch moieties. This group comprises polymers made from the fermentation of starch as well as mixtures of starch and natural or manufactured plastics. This makes up many of the thermoplastics already in use and represents around 50% of the worldwide bioplastics market such as thermoplastic starch (TPS) and Bio-PET.

ii. Bioplastics made of cellulose that is derived from cellulose esters or other cellulose derivatives. Because cellulose comprises glucose molecules linked together by a linkage [1, 14], certain symbiotic microorganisms are necessary for ruminants to digest it. For instance, cellulose acetate and methylcellulose.

iii. Aliphatic Polyesters—Materials that have more resistance to hydrolytic degradation, e.g., PHA and PLA.

iv. Protein-Based Bioplastics—Derivative of sources such as milk, wheat gluten, and other sources of protein. Very similar to the process of cheese-making, e.g., casein bioplastics.

v. Lignin-Based Bioplastics—Although lignin has long been a byproduct of cellulose manufacturing, it has only recently become important due to the development of biorefinery projects. For instance, PP- and PHA- and lignin polymer blends.

vi. Chitin-based bioplastic—The second most prevalent biopolymer after cellulose, chitin is comprised of N-acetyl-D-glucosamine units connected by linkages [1, 14]. Although chitin is found in the exoskeletons of arthropods and the cell walls of yeast and fungi, the shells of crustaceans like crabs, prawns, and shrimps are the main source of its extraction. For instance, bioplastics made of chitosan, chitin blended with PP, etc. [3].

1.2 Sources of bioplastics

Microbial biopolymers are natural polymers that are produced and broken down by a variety of species; they do not harm the host and have some benefits over petroleum-based plastics [14].

Because of their potential for usage and quick destruction by microorganisms, especially bacteria, biopolymers are innovative and promising. Under stressful circumstances, these biopolymers build up in microbial cells as store resources [15].

Microbiologically synthesized PHAs have shown considerable potential for various applications in the fields of (i) pharmaceuticals: controlled release and drug delivery systems; (ii) agriculture: regulated discharge of pesticides, plant growth regulators and herbicides, and fertilizers; (iii) biofuel: methyl ester of 3-hydroxybutyrate and methyl esters of 3-hydroxyalkanoate (MCL) can be used as biofuels; (iv) medicine: PHAs can be used to create absorbable sutures, bone plates, surgical pins, films, and staples, bone marrow supports, tendon repair tools, ocular implant implants, skin substitutes, cardiac valves, tissues for cardiovascular use, vascular grafts, tissue engineering applications, nerve guides, adhesion barriers, etc.; **(v)** disposable: PHAs may be utilized in the production of razors, food trays, diapers, hygiene items, cutlery, cosmetic packaging, glasses, medical surgical clothes, furniture, carpets, packaging, bags, compostable lids, and other items; and **(vi)** chromatography—Additionally, PHAs may be used as a stationary phase for chromatographic columns [16].

2. Polyhydroxyalkanoates

Bio-polymers such as polyhydroxyalkanoates (PHAs) are produced by microorganisms as lipid inclusions for granular types of energy storage inside the cellular structure [17]. PHAs are natural polyesters made from thermoplastic 3-, 4-, 5-, and 6-hydroxy alkanoic acids. More than 90 genera of bacteria, both Gram-positive and Gram-negative, have been found to produce PHAs in both aerobic and anaerobic conditions thus far. Some native bacterial strains, recombinant bacterial strains, and recombinant eukaryotes can all manufacture polyhydroxyalkanoates (PHAs). These bio polyesters are created by metabolically converting different carbon sources.

Numerous PHA polymers also offer intriguing characteristics, such as the ability to biodegrade, and they can be used for a variety of purposes, from single-use bulk plastics to specialized medicinal applications [18].

2.1 Structure of PHA

A total of 150 distinct PHA congeners have been identified. The resulting polymer is known as polyhydroxybutyrate or polyhydroxybutyric acid if the group is R = CH_3, polyhydroxyoctanoate (PHO) if R = C_3H_7, and so on.

2.2 Classification of PHA

PHAs are classified into three classes short, medium, or long chain length (scl, mcl, and lcl), respectively. It is based on the number of carbon atoms as short-chain-length PHA (scl-PHA), medium-chain-length PHA (mcl-PHA), and long-chain length PHA (lcl-PHA). Scl-PHA refers to PHA comprised of monomers having 5 or fewer carbon atoms [19]. These include 3-hydroxybutyrate and 3-hydroxyvalerate. The mcl-PHA is comprised of monomers having 6 to 14 carbon atoms. These include 3-hydroxyhexanoate, 3-octanoate, and 3-hydroxydecanoate. The lcl-PHA, which is uncommon and least studied, consists of monomers with more than 14 carbon atoms [20] (**Figure 1**).

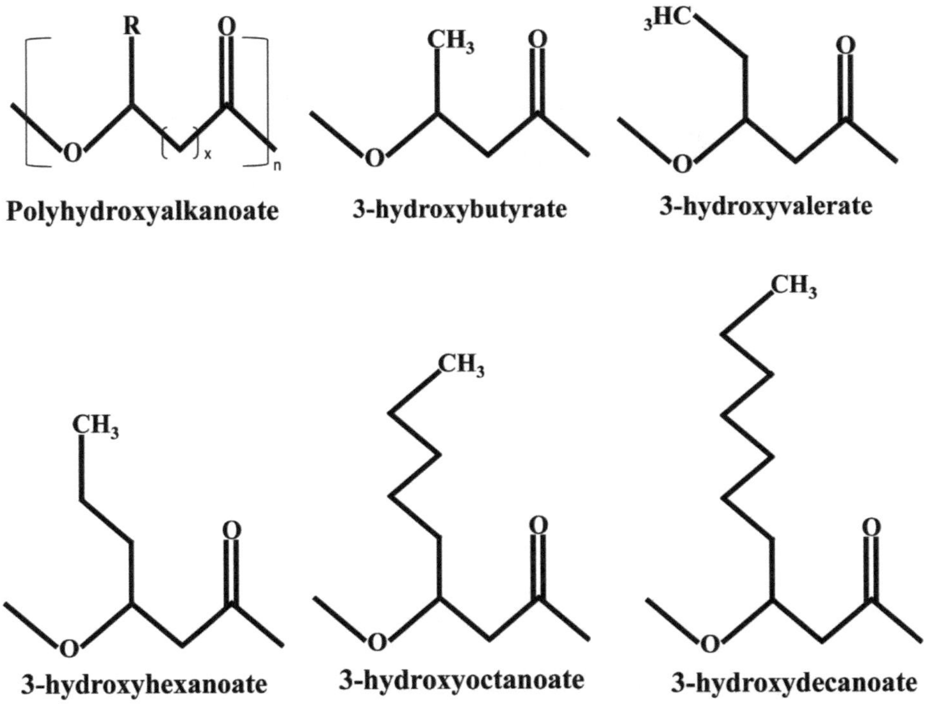

Figure 1.
General structures of polyhydroxyalkanoates.

2.3 Biosynthesis of PHA

PHAs are produced from two molecules of acetyl-CoA by three enzymatic reactions. The classical polyhydroxybutyrate (PHB) biosynthesis pathway consists of the following reactions:

1. β-ketothiolase (encoded by the phaA gene) catalyzes the formation of acetoacetyl-CoA by the condensation of 2 acetyl-CoA molecules.

2. Acetoacetyl-CoA dehydrogenase reduces acetoacetyl-CoA to R-3-hydroxybutyryl-CoA with the reduction of NADP(H) to $NADP^+$ (encoded by phaB gene).

3. PHA synthase polymerizes R-3-hydroxybutyryl-CoA to 3-hydroxyacid units (3HAs) or polyhydroxybutyrate (PHB) polymer (encoded by phaC gene).

However, apart from the classical pathway, there are other biosynthetic pathways involved in PHA production that differs based on the substrates, enzymes, and microorganisms used. The enzyme PHA synthase plays the most crucial role in PHA synthesis since it can polymerize 3-HA units obtained from different pathways such as fatty acid β-oxidation pathway, methylmalonyl-CoA pathway, and de novo fatty acid synthetic pathway [21, 22]. Numerous studies conducted on heterophilic bacteria revealed the classification of PHA synthase based on the specificity of 3-HA (C-Chain Length) substrate, amino acid sequence, and constituent subunits to have four classes [23]. Class I PHA synthases are encoded by phaC and polymerize scl-3HA units, monomers with approx 64 kDa MW. Class II PHA synthases polymerize mcl-3HA and are also encoded by phaC genes. These are monomers and have similar MW of ~63 kDa. Class III PHA synthases are heteromeric with ~40 kDa two subunits encoded by phaC and phaE genes each. They polymerize scl-3HA units. Class IV PHA synthases are similar to Class III and are encoded by either phaEC genes or phaRC genes. They polymerize scl-3HA to mcl-3HA and scl-3HA alone, respectively.

The acetyl-CoA utilized in the classical pathway of PHB synthesis is acquired as precursors derived from the tricarboxylic acid (TCA) cycle. This type of pathway is most commonly found in cyanobacteria, archaea, and heterophilic bacteria such as *Cupriavidus metallidurans*. Lipid metabolism is also used for the production of PHA which are mostly medium chain length (MCL) —PHAs. Different hydroxyalkanoates are generated from the β-oxidation pathway of fatty acids by the biotransformation of alkanes, alkenes, and alkanoates. The conversion of the β-oxidation intermediate trans-2-enoyl-CoA into (R)-hydroxyacyl-CoA is catalyzed by an R-specific enoyl-CoA hydratase (encoded by phaJ gene) and is the crucial step in this type of pathway. Studies conducted on *Aeromonas caviae* and *Pseudomonas putida* strains reported the (R)-specific manner of action of the phaJ enzyme [24, 25]. The PHA synthase (encoded by phaC genes) polymerizes (R)-hydroxyacyl-CoA into PHAs. MCL -3HA is produced in this type of pathway where both sugars and lipids are utilized. Glycolic precursor and fatty acid biosynthesis intermediates are converted to 3-hydroxyacyl-ACP by 3-hydroxyacyl-ACP-CoA transferase and malonyl-Coa-ACP transacylase. These key enzymes are encoded by phaG gene and are (R)-specific reactions by acyl-ACP-CoA transacylase. The 3-hydroxyacyl-ACP is converted into 3-hydroxyacyl-CoA and then polymerized to PHAs by PHA synthase.

Apart from the biosynthesis pathways and carbon source, other nutrients such as phosphate, nitrogen, oxygen, and sulfur also play a major role in PHA accumulation [26]. Limiting nitrogen and/or phosphorus with an excess carbon source is favorable for cell growth, along with C: N ratio changes showing better beneficial stress for PHA accumulation [15, 27, 28]. Under nitrogen deprivation, the conversion of α-ketoglutarate to glutamate is decreased causing accumulation of NAD(P)H by absorption of ammonium ions into cells. Similarly, the supplement of citrate reduces citrate synthase activity, thereby increasing the concentration of NAD(P)H. These high concentrations of NADPH result in increased PHB production since the reduction of acetoacetyl-CoA to R-3-hydroxybutyryl-CoA is increased [29]. Limiting phosphorus to a minimum level needed for cell maintenance restricts the Krebs cycle by promoting NADH accumulation, inhibiting citrate synthase and isocitrate dehydrogenase with increased acetyl-CoA. Nutritional stress induced by phosphorus limitation is sometimes more significant than nitrogen as a limiting factor in cyanobacteria and proved to be a good strategy for inducing PHA production [29] (**Figure 2**).

2.4 General properties of PHA

PHA properties are very indistinguishable from that of conventional plastics since it has great chemical diversity of radicals [17]. The ranges of these polymers vary from rigid and brittle thermoplastics to elastomers, rubbers, and adhesives which is totally based on their composition. Depending on the kinds of aromatic monomers used, aromatic PHAs exhibit a variety of properties. A lot of research has been done on the thermal characteristics of aromatic PHAs, which show behavior that is particular to the structure. Due to the increase in chain length and increase in the number of comonomers in a copolymer, its elasticity increases, and thus, PHAs have different properties according to their monomeric composition.

The physical properties of PHAs are as follows:

1. molecular mass

2. thermal properties

3. crystallinity index.

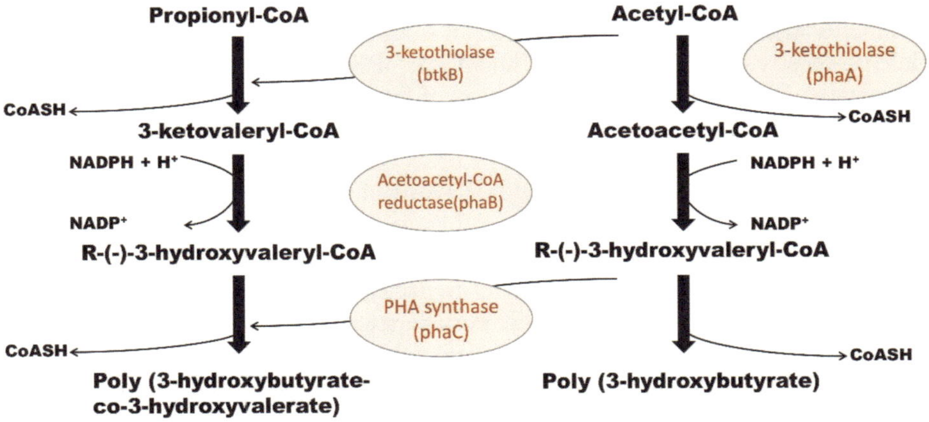

Figure 2.
Biosynthesis pathway of PHB and P(3HB-co-3 HV) copolymer. Adopted from [30].

Commercial suitability of molecular mass and molecular weight distribution of a polymer plays a vital role in characterization., and polymers with molecular mass less than 4×10^4 Da have their mechanical properties deteriorated.

The molecular weight of the compounds differs from 2×10^5 to 3×10^5 Da which depends on the type of microbial species used and growth conditions like pH, cultivation modes, and type and concentration of the carbon source. The properties of the PHA depend on the size of the polymer chains, whose structural rearrangements may depend on the degree of polymerization [31].

In addition to defining some mechanical characteristics of a material at ambient temperature, a polymer's thermal properties, such as its melting and glass transition temperatures as well as crystallinity and crystallization time, also serve as useful factors for the thermal processing of materials [32]. PHAs have melting points between 50 and 180°C and crystallinities between 30 and 70%, depending on the polymer's composition. PHAs are categorized as stiff if their crystallinity is between 60 and 80%. Medium (30–40%) and short (30%) polymer lengths characterize flexible and more elastic PHAs, respectively [31]. PHA's industrial applications are expanded thanks to its reduced degree of crystallinity, which also enhances its processing properties [32].

Semicrystalline polymers, the most popular type of PHA, are more brittle and less solvent-resistant but have tensile qualities that are comparable to those of polypropylene and polyethylene. PHB and its copolymers, which are made by cyanobacteria, have physical characteristics that can be linked to those of synthetic polymers like polypropylene and high-density polyethylene [33]. The creation of polymers with the appropriate properties will be aided by a good understanding of the connections between the PHA crystallinity and the polymer composition.

PHA is a suitable substitute for synthetic polymers due to its natural origin, biodegradability, biocompatibility, piezoelectricity, optical purity, and thermoplasticity [34]. Additionally, they are thermoplastic and/or elastomeric, non-toxic, and have a very high purity inside the cell. They are also hydrophobic, insoluble in water, inert, and indefinitely stable in the air [35]. PHA is less solvent resistant than polypropylene but has a substantially higher resilience to ultraviolet (UV) radiation degradation [36].

Numerous microorganisms in distinct situations have the ability to break down PHAs. PHA breakdown generates carbon dioxide and water under aerobic settings, whereas it generates carbon dioxide and methane under anaerobic ones [37]. The degradation time depends on a number of variables, including surface area, microbial activity of the environment, pH, temperature, humidity, the presence of other nutrients, and the properties of the polymer, such as composition and crystallinity, and can range from months (anaerobic digestion) to years (marine environment), among others.

Due to their high density, PHAs do not float in aquatic settings; as a result, after being dumped there, they sink and are biogeochemically destroyed on the surface of the sediments [37]. The two main processes involved in the biodegradation of polymeric heterocomposites, such as cellulose, starch, and aliphatic polyesters, of which PHAs are typical, are biotic or abiotic hydrolysis followed by bio-assimilation (hydrobiodegradation), and the second is peroxidation followed by the bio assimilation of low molecular mass (oxybiodegradation) products, which is applied in particular. Despite their quick biodegradability, PHAs are exceedingly stable in the air and do not decay when stored normally.

2.4.1 Appearance

Depending on the types of integrated monomers, aromatic PHAs have a variety of physical appearances. PHAs made only of phenoxy or phenyl monomers (P(3H5PhV)) are sticky and supple. When the content of 3H5PhV was increased in the instance of P(3HA-3H5PhV), the polymer softened. P(3HA-3-hydroxy-phenylalkanoate) [P(3HA-3HPhA)] changed from water-soluble to glue-like as the provided acyl chain length of phenylalkanoic acid was lengthened. PHAs with methylphenoxy groups are brittle, whitish substances [38]. PHAs that contain the 3H4BzB unit are similarly difficult. PHAs with thiophenoxygroups, however, are cream in color and elastomeric. The majority of PHAs that include the difluorophenoxy monomer is also cream-colored. Even with the addition of a small number of nitrophenyl units (1.2–6.9%), the physical properties of PHAs containing the nitrophenyl group diverged significantly from those of mcl-PHA [39].

2.4.2 Mechanical properties

The P(3-hydroxydodecanoate-3H5PhV) [P(3HDD-3H5PhV)] with varied 3H5PhV contents have different mechanical characteristics. The yield strength, maximum tension strength, and elongation at the break all decreased as a result of the addition of the 3H5PhV unit to P(3HDD). It is interesting to note that P(3HDD-18.70 mol% 3H5PhV) displayed a larger elongation at break than P(3HDD). On the other hand, except for P(3HDD-31.97 mol% 3H5PhV), Young's modulus increased above that of P (3HDD). These findings suggest a nonlinear relationship between the mechanical characteristics and the content of 3H5PhV [38].

2.4.3 Surface properties

Two fluorine atoms were added to P(3H5opFPxV), and its surface characteristics were assessed. This polymer has a surface contact angle of 104°, compared to 50° for PHAs having phenoxy or alkyl groups (C3 and C5) in the side chain [38]. A surface contact angle of more than 100 is typically insufficient to use the polymer as a non-wetting material. This difluorinated PHA thus possessed water-shedding qualities [39].

2.4.4 Degradability

The capacity of aromatic PHAs to degrade has also been investigated. One crucial quality of using PHAs as biodegradable materials is degradability. The stability of PHAs at physiological pH and the safety of the substance produced during hydrolysis should be assessed for medical applications such as medication delivery systems by analyzing the chemical degradation and microorganism-mediated degradation [38].

2.4.5 Chemical degradation

According to the literature, the P(3H6PhHx) homopolymer's chemical degradation was investigated. Around pH 7, this polymer is remarkably stable. It could therefore be utilized as a medication carrier to induce a delayed release of the active ingredient [39]. Additionally, the hydrolytic products of P(3H6PhHx) may have significant pharmacological effects that could enhance or expand the therapeutic effects of the drug that is encapsulated. These hydrolytic products can be oxidized in vivo to

phenylbutyric acid, phenylacetic acid, or trans-cinnamic acid. The antibacterial activity of (R)-3-hydroxy-phenylalkanoates (C5-C8), a hydrolytic product of PHAs bearing a phenyl group, is established. The relevant study showed that all (R)-3-hydroxy-phenylalkanoates inhibited the growth of Listeria species, attributed only (or mainly) to the phenyl group. Olivera et al. created polymeric microspheres of P(3H6PhHx) [38].

2.4.6 Solubility

Bacterial PHA copolymers often display a wide range of comonomer compositions, which may result from modifications in the bacterial metabolism during PHA production. The biosynthesized aromatic PHAs are not always formed as a copolymer, but rather occasionally as a combination of two distinct PHAs. These aromatic polymers were isolated by solvent fractionation in several investigations [38].

2.4.7 Thermal properties

PHAs are polymers that are only partly crystalline. Therefore, the Tg and Tm of the amorphous and crystalline phases are typically used to express the thermal characteristics of these materials. The results of several studies show that the properties of aromatic PHAs differ significantly from those of mcl-PHAs, which are elastomers with Tgs between 53 and 28 C and a Tm between 45 and 69 C, where the values change depending on the types of aromatic monomers used [38].

2.4.8 Extraction of PHA

Treatment of cellular disruption and/or instability, recovery, and purification of biopolymers are the steps involved in the PHA extraction process. These procedures allow for the use of chemical, physical, biological, or even a mix of these technologies to provide a product with high purity and preserved physical and thermal characteristics.

The first step in the PHA extraction method is to centrifugate the solid material, which is made up of cells containing intracellular biopolymer, from the culture broth. Additionally, the microbial cell wall may be punctured or disturbed through biological, physical, or chemical means [40]. A suspension of bio-polymer, cells containing biopolymer (cells that destabilize but do not break cell walls), and cell debris form upon rupture or instability of the cell wall (mixture of proteins, nucleic acids, lipids, and cell-wall fragments). The next stage is to recover the biopolymer, which can be done in a variety of ways, including chemically, biologically, physically, or utilizing a combination of approaches like physical and chemical, biologically and chemical, among others [16].

2.4.9 Chemical methods

Isolated or coupled solvents are used in the chemical processes of removing PHAs from the cells of the microorganisms [40]. Chloroform, acetone, methyl isobutyl ketone, methylene chloride, propylene carbonate, ethyl acetate, and isoamyl alcohol are the most often used solvents. It is vital to assess the contact time and heating temperature of the polymer with the solvent to gauge the efficacy of the extraction process and the quality of the resulting product [16].

2.4.10 Physical methods

The use of homogenizer mills and ultrasound is among the most popular physical techniques used in PHA extraction. These methods are typically used at the beginning of the extraction procedure to disrupt and weaken the microorganisms' cell membranes. When compared to chemical extraction techniques, mechanical extraction yields polymers with higher thermal characteristics while also being more cost-effective and less hazardous. If an appropriate chemical method is used in conjunction with the mechanical method to extract biopolymers, which allows for high PHA recovery without significantly altering its features, the possibility for recovery will be increased [16].

2.4.11 Biological methods

The biological method of microbial PHA extraction is a complicated procedure that relies on the use of enzymes including lysozymes, nucleases, and proteases to recover the biopolymer. The culture broth is supplemented with enzymes to hydrolyze PHA-containing cells [16]. The gentle operating conditions, great selectivity of the enzymes in hydrolyzing the microorganisms' cell wall proteins without affecting the breakdown of the polymer, and high quality of the recovered polymer make this technology appealing [40].

3. Cyanobacteria as a source of bioplastics

The PHB accumulation in cyanobacteria was first reported by Carr G.N. in 1966 with up to 10% (dcw) in *Chlorogloea fritschii* [41]. In photoautotrophic culture *Artrospira platensis (Spirulina)* accumulated a maximum PHB of 6%, and it is very less for exploiting cyanobacteria for PHA thermoplastics production. However, PHA biosynthesis in, *Synechocystis sp.* PCC 6803, *Nostoc muscorum*, and *Synechococcus sp.* MA19 produced up to 38, 46, and 55% (dcw), respectively, under different limiting culture conditions reported in studies [9, 42, 43]. The production of PHA reported in several other strains in photoautotrophic and also with supplementation of acetate or other organic carbon sources are profoundly lower compared to heterotrophic bacteria. Chen et al. reported a maximum accumulation of poly(3-hydroxybutyrate-co-3-hydroxyhexanoate) [P(3HBco-3HHx)] co-polymer up to 50% (dcw) in *Aeromonas hydrophila* 4AK4 grown in 5% glucose medium with 5% lauric acid under phosphorous limitation with a productivity of 540 $mg.L^{-1}.h^{-1}$ [44]. Despite the high PHA accumulation, bacterial PHA thermoplastic has commercial limitations since the organic carbon substrate itself accounts for ~30–50% of the total cost of production on a large-scale [15]. For example, PHB production of up to 77% (dcw) was reported in recombinant *Escherichia Coli* using glucose as substrate with a productivity of 3200 $mg.L^{-1}.h^{-1}$; however, the carbon source used accounts for 38% of the overall cost of production [45]. Compared to heterotrophic bacteria (4–5% carbon substrate), cyanobacteria required significantly lower carbon substrate at about 0.4% [44, 46]. Thus, cyanobacteria are a more promising candidate for the large-scale production of bioplastics.

3.1 General Cultivation of cyanobacteria

Cyanobacteria are photosynthetic prokaryotes found in both fresh and marine water, soil, etc., and they have a unique physiology that makes them survive even in

harsh ecological habitats such as deserts, hot springs, volcanic substrates, and even in alkaline basins. The cyanobacteria can be cultivated in three different culture systems—an open-raceway pond (mostly preferred), a closed system (photobioreactor), and a hybrid system (combination of both open and closed systems) [47]. The widely used media for the cultivation of cyanobacteria is BG11 having the following composition (1500 $mg.L^{-1}$ $NaNO_3$, 31.4 $mg.L^{-1}$ K_2HPO_4, 36 $mg.L^{-1}$ $MgSO_4$, 36.7 $mg.L^{-1}$ $CaCl_2.2H_2O$, 20 $mg.L^{-1}$ Na_2CO_3, 1 $mg.L^{-1}$ NaMgEDTA, 5.6 $mg.L^{-1}$ citric acid, 6 $mg.L^{-1}$ ferric ammonium citrate, and 120 mg·L-1 $NaHCO_3$) (himedialabs). The components of the modified BG-11 used in the reactors are (K_2HPO_4, $NaNO_3$, $NaHCO_3$, $CaCl_2.2H_2O$, NaOH, Na_2EDTA, and NaHCO3). The optimum pH and temperature for the growth of cyanobacteria are 7.5–9 and 30 $\pm$ 2°C, respectively. The culture takes up to 7 days to reach the log phase, and the complete growth cycle ends in 20 days (after reaching the death phase).

3.1.1 Open systems

Open ponds are the natural ecosystem in which the algae tend to grow and develop. Open systems are classified into two types—natural (lakes and ponds) and artificial (containers and artificial ponds). There are several advantages of growing cyanobacterium in open systems which include low investment, construction of the pond being easier, and easy maintenance. Some of the drawbacks include a requirement for large land, poor light penetration, and low biomass productivity [48].

3.1.2 Closed systems

Photobioreactors are considered to be the closed system for the cultivation of cyanobacterium. By using this culture system, the drawbacks of the open system can be neglected. There are several advantages of using a closed system for algal cultivation which include control over culture parameters (pH, temperature, etc.,), low level of contamination, and good mixing that induces high gas exchange within the culture. There are various types of closed system available for the culture of algae which includes vertical column, tubular bioreactor, flat-plate bioreactor, etc [48].

3.1.3 Hybrid system

A combination of both open and closed systems is known as a hybrid system. There are two stages of cultivation in which the first stage involves a closed system and the second stage occurs in the open-raceway system. By utilizing this system, the advantages of both open and closed systems are possible. Many ongoing studies are designing a commercial-scale hybrid reactor that can be economical and can be easy to handle [49].

4. PHA from cyanobacteria

4.1 Biosynthesis of PHA in cyanobacteria

For decades it was believed that cyanobacteria possess an incomplete Kerbs cycle like some prokaryotes due to the absence of the 2-oxoglutarate dehydrogenase

complex which performs the conversion of 2-oxoglutarate to succinyl-CoA in the TCA cycle [50]. Since the TCA cycle is incomplete, it is assumed that the breakdown of PHB polymers generating acetyl-CoA could be utilized neither for the production of cell components nor for energy generation [51]. It was hypothesized that this cycle was closed by the glyoxylate stunt of aspartate transaminase reactions [52]. However, recent studies reported that the Kerbs cycle was completed with help of γ-aminobutyric acid shunt and 2 enzymes 2- oxoglutarate decarboxylase and succinic semialdehyde dehydrogenase found in *Synechocystis sp.* PCC 6803 [53] and *Synechococcus sp.* PCC 7002 [54], respectively. The later reported protein-encoding genes are present in most cyanobacteria with variation in their organization.

The PHA polymer biosynthesis is linked with mobilization or depolymerization [5]. The PHA polymers usually undergo a cyclic process of biosynthesis and depolymerization, where the PHA is formed from acyl-CoA precursors via different metabolic routes under nutrient depletion/limitation conditions as the carbon source is stored as polymer granules in the cells. The mobilization of PHB polymers is carried out by intracellular PHB depolymerase generating acetyl-CoA which is used to generate oxidation via the Krebs cycle. Many studies reported the regulatory effect of acetyl phosphate produced by the phosphotransacetylase catalytic activity on the post-translation of PHB synthase enzyme [55–59]. The exploitation of exogenous carbon sources such as glucose, fructose, and acetate showed decreased mobilization and increased biosynthesis of PHA [60–63].

4.2 PHA production

The PHA-producing cyanobacterium is classified into two groups—one group requires a limitation of an essential media component for PHA production, and another group does not require any limitation in nutrients for the production of PHA. The cyanobacterium that can be cultivated without nutrient limitation is preferred on an industrial scale. A few studies have been conducted to optimize the nutrients for the production of PHA and PHB on large scale in batch mode. In a study, *Synechocystis* sp. PCC 6803 was cultivated in BG11 media with reduced nitrogen concentration and showed a maximum PHB accumulative of 180 $mg.ml^{-1}$ [47], *Synechocystis* sp. CCALA192 was cultivated in a 200 L tubular photobioreactor in batch mode and accumulated a maximum of 125 $mg.ml^{-1}$ of PHB, and a wild-type cyanobacterial strain *Synechocystis* sp. PCC 6714 produced a maximum of 640 $mg.L^{-1}$ of PHA when cultivated in optimized growth media [64].

Several studies reported that higher PHA accumulation in cyanobacteria occurs under nutritional stress activating the PHA biosynthesis pathway. According to Mendhulkar and Shetye [65], the metabolic pathways are diverted to produce carbon-rich compounds for energy storage, such as PHAs, and glycogen, when the cyanobacteria experience nutrient deficiency (nitrogen and/or phosphorus). The study on cyanobacteria *Synechococcus subsalsus* and *Spirulina sp.* LEB18 in nitrogen-deficient environment revealed that the carbon source is diverted to other metabolic pathways for biopolymer production which is used as energy storage and reused in favorable conditions [66]. PHA accumulation in *Botryococcus braunii* and *Synechocystis salina* grown in BG-11 medium without any nutritional limitation was reported [67, 68]. Different nutritional conditions are employed to increase the production of PHA such as excess or limited levels of nitrogen and/or phosphorus, acetate, and propionate, and various other conditions like salinity, gas exchange, wastewater as a source, etc., were summarized in (**Table 1**). Apart from culture condition variations,

Cyanobacteria	PHB content (% DCW)	Substrate	Production condition	Polymer composition	Reference
Synechocystis sp. PCC 6803	38	Acetate	P limitation and gas exchange limitation	PHB	[69]
Synechocystis sp.	11		Nitrogen and phosphorous deficiency	PHB	[69]
Synechococcus sp.MA19	55		Phosphorous deficiency	PHB	[43]
Synechocystis sp. PCC 6714	16	CO_2	N^2 and P^3 limitation	PHB	[12]
Spirulina platensis	6.0	CO_2	Not given	PHB	[70]
Spirulina platensis UMACC 161	10	Acetate and CO_2	N starvation	PHB	[71]
Botryococuus braunli	16.4	Sewage wastewater	BG 11 medium	PHB	[67]
Spirulina sp. LEB-18	12		Nitrogen deficiency	—	[66]
Spirulina platensis	10		Addition of acetate and CO_2	PHB	[71]
Synechocystis salina	5.5–6.6	CO_2	BG 11 medium	PHB	[68]
Synechococcus subsalsus	16		Nitrogen deficiency	—	[66]
Spirulina maxima	7–9	CO_2	N and P limitation	PHB	[72]
Synechocystis sp. PCC6803	5		BG 11 medium	—	[73]
Synechococcus elongates	17.15	Sucrose	Nitrogen deficiency	PHA	[65]
Synechococcus elongates	7.02	Sucrose	Phosphorous deficiency	—	[65]
Gloeothece sp. PCC 6909	9.0	Acetate	—	—	[15]
Microalgae consortium	43	Agro-based industrial wastewater and activated sludge	Wastewater	PHB	[74]
Microalgae consortium	31	Agro-based industrial wastewater and activated sludge	Wastewater	PHB	[75]
Nostoc muscorum	69		Phosphorous deficiency	P(3HB-co-3 HV)	[10]
N. muscorum	31	Acetate and propionate	Addition of acetate and propionate	P(3HB-co-3 HV)	[9]

Cyanobacteria	PHB content (% DCW)	Substrate	Production condition	Polymer composition	Reference
N. muscorum Agardh	60	Acetate and valerate	N deficiency	PHB-co-PHV	[11]
N. muscorum	22	CO_2	P starvation	PHB	[61]
Spirulina subsalsa	7.45	Acetate and CO_2	Increased salinity	PHB	[76]
Spirulina sp. LEB18	30.7		Nitrogen deficiency	PHB	[77].
Aulosira fertilissima	49	Acetate	Gas exchange limitation	PHB	[78]
Alusira fertilisim CCC444	77	Fructose and valerate	N deficiency	PHB-co-PHV	[79]
Alusira fertilisima CCC444	85	Citrate and acetate	P deficiency	PHB	[80]
Synechocystis PCC 7942	3	CO_2	N limitation	PHB	[81]
Synechocystis PCC 7942	25.6	Acetate	N limitation	PHB	[81]
Synechocystis sp. CCALA192	12.5	CO_2	N limitation	PHB	[13]
Anabaena cylindrica	<0.005	CO_2	Balanced Growth	PHB	[82]
A. cylindrica	2.0	Propionate	N limitation	PHB + PHV	[82]
Synechococcus elongatus	17.2	CO_2 and sucrose	N deficiency	—	[65]
Caltorix scytonemicola TISTR 8095	25	CO_2	N deficiency	PHB	[83]

Table 1.
PHA production in cyanobacteria under different culture conditions.

highly productive strain selection can also increase the PHA accumulation yields ranging from 5.0% to about 70% (dcw).

Coelho et al., [77] reported higher percentages of PHA accumulation in *Spirulina sp.* using Zarrouk medium with nitrogen and phosphorus limitations of 30.7% and 14.1% (dcw), respectively. Phosphorus and gas exchange limitations along with additional acetate and nitrogen and phosphorus limitations in *Synechocystis sp.* PCC 6803 lead to PHA accumulation of about 38% and 11% (dcw), respectively [69]. Studies conducted by Bhati and Mallick on PHB-PHV co-polymer production in *N. muscorum* under nitrogen and phosphorus deficiency resulted in co-polymer accumulation of about 60% and 69% (dcw), respectively [10, 11]. Samantaray and Mallick reported a maximum of 85% (dcw) PHB and 77% (dcw) PHB-co-PHV in *Alusira fertilisima* CCC444 under nitrogen deficiency with fructose and valerate supplementation and phosphorus deficiency along with additional citrate and acetate, respectively [79, 80].

4.3 Strategies to improve cyanobacterial PHA production

4.3.1 Genetic manipulation

Many studies have been conducted on gene manipulation of cyanobacteria on metabolic engineering and PHB synthesis, *Synechhocystis sp.* PCC 6803 is the most studied strain. Insertion of *C. nectar* PHA operon into *Synechococcus* PCC 7942 increased the PHA production from 3 to 25% (dcw) [84]. The *Synechhocytis sp.* PCC 6803 was transfected with the PHA synthase gene from *C. nectar* and showed increased activity but net PHB content did not increase [73]. Overexpression of phaAB with 4 mM acetate supplementation showed an increase in PHB of up to 35% (dcw) in *Synechocytis sp. PCC6803* [85]. Wang et at. reported volumetric productivity of 263 $mg.L^{-1}.d^{-1}$ and a yield of 1.84 $g.L^{-1}$ by overexpression of the acetoacetyl-CoA reductase gene in *Synechocystis* [86]. **Table 2** summarizes further studies conducted on genetic manipulation for increasing PHB production.

4.3.2 Suppressing glycogen synthesis pathway

The 3PG intermediate is utilized for both glycogen and PHB polymer production. The productivity of glycogen is high and quicker than that of PHB in nitrogen deprivation conditions (30% PHB and 60% glycogen (dcw) is produced) [92]. Assimilation of CO_2 through ribulose-1,5-biphosphate carboxylation by the Rubisco produces 3PG which is directed to glycogen biosynthesis more than PHB accumulation. Grundel

Cyanobacteria	Genetic manipulation	Culture conditions	PHB content (% DCW)	Reference
Synechocystis sp. PCC 6803	Overexpression of PHA synthase	Direct photosynthesis	14	[87]
Synechocystis sp. PCC 6803	Transconjugant cells harboring expression vectors carrying PHA genes	CO_2	7.0	[88]
Synechocystis sp. PCC 6803	Introducing PHA biosynthetic genes from C. nectar	Acetate and nitrogen limitation	11	[73]
Synechocystis sp. PCC 6803	Increasing acetyl-CoA levels	CO_2	12	[89]
Synechocystis sp. PCC 6803	Overexpression of native PHA genes	CO_2 and nitrogen deprivation	26	[85]
Synechocystis sp.	Optimization of acetoacetyl-CoA reductase binding site	CO_2	35	[86]
Synechococcus sp. PCC 7942	Defective in glycogen synthesis	CO_2	1.0	[90]
Synechococcus sp. PCC 7942	Introducing PHA biosynthetic genes from C. nectar	Acetate and nitrogen limitation	26	[81]
Synechococcus sp. PCC 7002	Introduction of GABA Shunt	CO_2	4.5	[91]

Table 2.
Genetic manipulations to increase PHB biosynthesis.

et al. reported that there is no influence on growth under continuous light conditions while the biosynthesis pathway of glycogen was impaired in *Synechocystis sp.* PCC 6803 [93]. In the study conducted by Wu et. al, [94], an increase in PHB accumulation from 8–13% was observed in knockout mutants unable to produce glycogen and did not turn into dormant mode and was unable to recover from nitrogen scarcity. However, PHB-deficient mutants produced the same level of glycogen as the wild-type and recovered from scarcity once replenished with nutrients. A deficiency of growth was observed in the mutants with the knockout of genes involved in both polymer syntheses. Thus, it is important to improve the synthesis of PHB yield with robust PHB production and suppressed glycogen pathway.

4.3.3 Exploitation of metabolic inhibitors to increase cyanobacterial PHA

As the imbalance of C: N and NADPH: ATP ratios are contributing factors in stimulating PHB production many studies were carried out on the effect of the metabolic inhibitor on PHB production. Upon supplementing *N. muscorum* with carbonylcyanide m-chlorophenylhydrazone (CCCP) and dicyclohexylcarbodiimide (DCCD), the PHB pool was increased to 21% and 17% from 8.5%, respectively, were reported [55]. The addition of monofluoroacetate increased the PHB pool up to 19% (dcw), while Lmethionine-DL-sulfoximine (MSX) and azaserine addition also enhanced PHB production. Treatment with metabolic inhibitors such as DCCD, CCCP, and [3-(3,4-di chlorophenyl)-1,1- dimethylurea influenced the NADPH: NADP ratio along with PHB accumulation in *Synechocystis* PCC 6803 were reported [95]. This strategy of using metabolic inhibitors could help to enhance PHA accumulation in both wild-type and recombinant cyanobacteria.

4.3.4 Mixed consortium

The mixed consortium of cyanobacteria, bacteria, and algae is a feast-famine strategy where a sequencing batch reactor (SBE) without aeration is used for the cultivation [96, 97]. The concept of the consortium is developed to increase the system efficiency by enhancing productivity and accessibility of resources, community stability, efficient nutrient cycling, and partitioning, and distribution of carbon or energy source in a non-competitive manner. The oxygen produced by the algae cells during the famine phase is used to consume the NADPH reserves of the cells leading to around a 20% (dcw) increase in PHA accumulation [97]. A permanent feast regime under high light intensity conditions promoted PHA production to a maximum of 60% (dcw) in photosynthesis mixed culture. The famine phase can be eliminated using axenic dark feast conditions increasing productivity by up to 60% (dcw) by facilitating the acetate uptake [96, 97].

4.3.5 Two-Stage cultivation

The two-stage cultivation strategy is exploited for high biomass production and increased concentration of PHA thermoplastics. The cells are grown in optimal nutritional conditions in the first or growth stage to achieve high biomass concentration. The cells are recultivated in fresh media with the limitation of a specific nutrient (nitrogen and/or phosphorous) in the second or accumulation stage to induce stress and produce PHA. A study conducted on *Chlorogloea fritschii* TISTR 8527 in two-stage cultivation shows a maximum PHB accumulation of 25% (dcw) using acetate as

substrate with 51 ± 7% (w/w) of conversion efficiency [83]. As the first stage produces maximum biomass this strategy appears to be potentially viable for large-scale production, but the shear forces experienced by the cells during recultivated give rise to a new lag phase. Two-stage cultivation of *Synechocystis cf. salina* PCC 6909 operated in a single stage without recultivation of biomass produced about 90 $mg.L^{-1}$ of PHAs in 14 days. The cyanobacterium was grown in an optimized media such that the phosphorous and nitrogen were almost utilized by 7–8 days with a maximum biomass production of up to 2 $g.L^{-1}$ (dcw) and thereby entered the accumulation stage due to nutrient starvation without harvesting and transfer of biomass [98]. The overall production cost of PHA production can be reduced using such type of two-stage cultivation strategy.

5. Problems

Currently, the major bottleneck is the non-existence of an economical mass cultivation strategy for the commercial production of cyanobacterial PHAs. The two commercial-scale mass cultivation approaches as **(i)** closed photobioreactors and **(ii)** conventional open pond culture systems. The close photobioreactors are effective for monoculture cultivation as they are of more controlled types. An ideal photobioreactor should be flexible to all system requirements for different strains and specific growth environments for the production of the product of interest [99]. Open pond culture system is cheaper compared to photobioreactor which requires high construction, operation, and maintenance cost.

Biomass harvesting from the water on a commercial scale is still a major issue partly due to the low concentration (0.2–2 $g.L^{-1}$), small size, and colloidal stability [100]. Filtration, flocculation, gravity settling, and centrifugation are some of the techniques exploited for harvesting cyanobacterial biomass. Flocculation is cost-effective and energy efficient compared to centrifugation and it can also handle a huge volume of culture. Addition of inorganic salts such as $AlCl_3$, $Al_2(SO_4)_3$, $FeCl_3$, and so on, cationic starch and chitosan are used for the flocculation of biomass [101, 102]. Several research efforts are being carried out for developing cost-effective and efficient cyanobacterial biomass harvesting technologies. For example, the settling velocity distribution of flocculated microalgal/cyanobacterial biomass is a critical parameter for developing cost-effective gravity settlers for biomass recovery.

The drying of biomass is essential for further downstream processing and storage. Around 20% of the overall cost of PHA production from *Spirulina* is contributed to the drying process. The high-energy input process of drying is only required for PHA extraction. Air drying is quite feasible, but it requires a large area and a longer time. Solar or wind energy utilization for the drying process could overcome these limitations [6].

6. Applications

PHA has a lot of advantages over conventional plastics because of its sustainability, now fossil plastics are to be replaced the major obstacle to be faced is the reduction of the cost associated with microbiological plastic production. The cost of producing traditional petroleum-based plastic in 2002 was €1.00/kg, which was considerably less than the €9.00/kg cost of PHA. Even when compared to other sustainable polymers, like PLA, which costs €1.72/kg, microbiological manufacture of PHA costs €2.49/kg,

which is still pricey [103, 104]. Carbon source plays an important role in facing the obstacles such as yield of the input, fermentation, productivity, and downstream processing [105, 106].

7. Strategies to choose to face the obstacle related to a circular economy and industrial ecosystem

Keen interest in cyanobacteria is because of the production of different metabolites which works with more than one type of compound as a salable product this type of application use is called a "cradle to cradle" system (turning waste into a new product) that is bioplastic [107]. Another instance of turning waste into a new product is using microalgae, reusing the effluents from the refining of olive oil in the cultivation of microalgae for biodiesel and biopolymers [29].

Another beneficial environmental effect that makes the adoption of a circular bioeconomy more real is the uptake of ambient carbon dioxide for conversion into biotechnological products. Using by-products and leftovers from microbiological production, it is possible to integrate the creation of bioplastic with the manufacture of other desirable goods to reduce the cost of microbial PHB. An effective alternative is the cyanobacterial genus *Nannochloropsis sp.,* which produces eicosapentaenoic acid, and the cyanobacterial genus *Spirulina platensis*, which produces linoleic acid. This species is important for its expressive biomass output, which has a high protein content and can be used to make animal feed or nutraceuticals.

The construction of a biorefinery, merging *Synechocystis salina's* PHB synthesis with commercially valuable pigments, notably the commonly abundant phycocyanin and chlorophyll, and carotenoids, showed encouraging results. Since the quality of the resulting polymer is directly influenced by purification, which includes the removal of pigments that can be employed in manufacturing chains of higher value, the extraction of pigments without their degradation is not only feasible but also necessary. In addition to pigments, *S. salina* biomass contains carbohydrates, lipids, and proteins that can be used as animal feed as long as the necessary nutritional standards and laws regarding the presence of contaminants like heavy metals or mycotoxins are observed. In this case, cyanotoxins are given priority over cyanobacteria that do not produce toxins [108].

Cyanobacterial dietary supplements are also advantageous for animal health, with *Spirulina sp.* biomass enhancing hens' humoral and immunological responses. For cyanobacteria and microalgae in general, the dual benefit of production connected with bioremediation has already been discussed, with a focus on the creation of biodiesel. The same idea can be used to explain how naturally transformable organisms like cyanobacteria can produce biopolymers, opening new opportunities for genetic engineering.

8. Conclusion

PHA has turned out to be a substitute for conventional plastics. Cyanobacteria is becoming the alternative source of PHA production. The major cause of the production of PHA using microalgae is to reduce the cost. Now, cyanobacteria aids in the production of PHA as it collects a huge amount of PHS through photosynthesis which ultimately requires less nutritional content for growth. Cyanobacteria have a very low

yield of autotrophic PHA; in near future, a biological system can be constructed to make use of the resources and attributes which can increase in production of PHA through autotrophic and heterotrophic. Production of PHA using microalgae has many other advantages like industrial compounds, which include pigments, antioxidants, cosmetics, pharmaceuticals, polysaccharides, and so on. Additionally, it has been noted that these organisms create a variety of secondary metabolites, poisons, and other bioactive substances that are significant from a pharmacological perspective. The economics of cyanobacterial PHAs would unquestionably be improved by integrating all of these substances under a refining method.

Author details

Mohanasundaram Yogeswar[1], Natarajan Valle[1] and Arumugam Nagarajan[2]*

1 Center for Biotechnology, Alagappa College of Technology, Anna University, Chennai, India

2 Applied and Industrial Microbiology Laboratory, Department of Biotechnology, Indian Institute of Technology Madras, Chennai, India

*Address all correspondence to: nagarajanarumugamm@gmail.com

References

[1] Ashter SA. Types of Biodegradable Polymers. In: Plastics Design Library, Introduction to Bioplastics Engineering. William Andrew Publishing; 2016:81-151. DOI: 10.1016/B978-0-323-39396-6.00005-1

[2] Singh R, Parihar P, Singh M, Bajguz A, Kumar J, Singh S, et al. Uncovering potential applications of cyanobacteria and algal metabolites in biology, agriculture and medicine: Current status and future prospects. Frontiers in Microbiology. 2017; **8**(APR):515

[3] Nandakumar A, Chuah JA, Sudesh K. Bioplastics: A boon or bane? Renewable and Sustainable Energy Reviews. 2021; **147**(May):111237. DOI: 10.1016/j.rser.2021.111237

[4] Venkata Mohan S, Hemalatha M, Chakraborty D, Chatterjee S, Ranadheer P, Kona R. Algal biorefinery models with self-sustainable closed loop approach: Trends and prospective for blue-bioeconomy. Bioresource Technology. 2020;**2020**:295. Available from: https://pubmed.ncbi.nlm.nih.gov/31563289/

[5] Anderson AJ, Dawes EA. Occurrence, metabolism, metabolic role, and industrial uses of bacterial polyhydroxyalkanoates. Microbiological Reviews. 1990;**54**(4):450-472. Available from: /pmc/articles/PMC372789/?report=abstract

[6] Singh AK, Mallick N. Advances in cyanobacterial polyhydroxyalkanoates production. FEMS Microbiology Letters. 2017;**364**(20):189. Available from: https://academic.oup.com/femsle/article/364/20/fnx189/4107776

[7] Koch M, Bruckmoser J, Scholl J, Hauf W, Rieger B, Forchhammer K. Maximizing PHB content in Synechocystis sp. PCC 6803: A new metabolic engineering strategy based on the regulator PirC. Microbial Cell Factories. 2020;**19**(1):1-12. Available from: https://microbialcellfactories.biomedcentral.com/articles/10.1186/s12934-020-01491-1

[8] Chavan S, Yadav B, Tyagi RD, Drogui P. A review on production of polyhydroxyalkanoate (PHA) biopolyesters by thermophilic microbes using waste feedstocks. Bioresource Technology. 2021;**341**:125900

[9] Sharma L, Kumar Singh A, Panda B, Mallick N. Process optimization for poly-β-hydroxybutyrate production in a nitrogen fixing cyanobacterium, Nostoc muscorum using response surface methodology. Bioresources Technology. 2022;**98**(5):987-993. Available from: https://pubmed.ncbi.nlm.nih.gov/16765593/

[10] Bhati R, Mallick N. Poly (3-hydroxybutyrate-co-3-hydroxyvalerate) copolymer production by the diazotrophic cyanobacterium Nostoc muscorum Agardh: Process optimization and polymer characterization. Algal Research. 2015;**7**:78-85

[11] Bhati R, Mallick N. Production and characterization of poly (3-hydroxybutyrate-co-3-hydroxyvalerate) co-polymer by a N2-fixing cyanobacterium, Nostoc muscorum Agardh. Journal of Chemical Technology Biotechnology. 2012;**87**(4):505-512. Available from: https://onlinelibrary.wiley.com/doi/full/10.1002/jctb.2737

[12] Kamravamanesh D, Pflügl S, Nischkauer W, Limbeck A, Lackner M, Herwig C. Photosynthetic poly-β-

hydroxybutyrate accumulation in unicellular cyanobacterium Synechocystis sp. AMB Express. 2017; 7(1):6714. Available from: https://pubmed.ncbi.nlm.nih.gov/28687036/

[13] Troschl C, Meixner K, Fritz I, Leitner K, Romero AP, Kovalcik A, et al. Pilot-scale production of poly-β-hydroxybutyrate with the cyanobacterium Synechocytis sp. CCALA192 in a non-sterile tubular photobioreactor. Algal Research. 2018;**34**:116-125

[14] Anjum A, Zuber M, Zia KM, Noreen A, Anjum MN, Tabasum S. Microbial production of polyhydroxyalkanoates (PHAs) and its copolymers: A review of recent advancements. International Journal of Biological Macromolecules. 2016;**89**: 161-174. Available from: https://pubmed.ncbi.nlm.nih.gov/27126172/

[15] Singh AK, Sharma L, Mallick N, Mala J. Progress and challenges in producing polyhydroxyalkanoate biopolymers from cyanobacteria. Journal of Applied Phycology. 2017;**29**(3): 1213-1232. DOI: 10.1007/s10811-016-1006-1

[16] Costa SS, Miranda AL, de Morais MG, Costa JAV, Druzian JI. Microalgae as source of polyhydroxyalkanoates (PHAs) — A review. International Journal of Biological Macromolecules. 2019;**131**: 536-547. DOI: 10.1016/j.ijbiomac.2019.03.099

[17] Poli A, Di Donato P, Abbamondi GR, Nicolaus B. Synthesis, production, and biotechnological applications of exopolysaccharides and polyhydroxyalkanoates by Archaea. Archaea. 2011;**2011**

[18] Lu J, Tappel RC, Nomura CT. Mini-review: Biosynthesis of poly (hydroxyalkanoates). Polymer Reviews. 2009;**49**(3):226-248

[19] Kunasundari B, Sudesh K. Isolation and recovery of microbial polyhydroxyalkanoates. Express Polymer Letters. 2011;**5**(7):620-634

[20] Rai R, Keshavarz T, Roether JA, Boccaccini AR, Roy I. Medium chain length polyhydroxyalkanoates, promising new biomedical materials for the future. Materials Science & Engineering R: Reports. 2011;**72**(3):29-47

[21] Singh M, Patel SKS, Kalia VC. Bacillus subtilis as potential producer for polyhydroxyalkanoates. Microbial Cell Factories. 2009;**9**(1):1-11. Available from: https://microbialcellfactories.biomedcentral.com/articles/10.1186/1475-2859-8-38

[22] Singh M, Kumar P, Ray S, Kalia VC. Challenges and opportunities for customizing Polyhydroxyalkanoates. Indian Journal of Microbiology. 2015; **55**(3):235-249. Available from: https://link.springer.com/article/10.1007/s12088-015-0528-6

[23] Rehm BHA. Polyester synthases: Natural catalysts for plastics. Biochemical Journal. 2003;**6**(1):15-33. Available from: https://pubmed.ncbi.nlm.nih.gov/12954080/

[24] Tsuge T, Fukui T, Matsusaki H, Taguchi S, Kobayashi G, Ishizaki A, et al. Molecular cloning of two (R)-specific enoyl-CoA hydratase genes from Pseudomonas aeruginosa and their use for polyhydroxyalkanoate synthesis. FEMS Microbiology Letters. 2000; **184**(2):193-198. Available from: https://pubmed.ncbi.nlm.nih.gov/10713420/

[25] Fukui T, Doi Y. Cloning and analysis of the poly(3-hydroxybutyrate-co-3-hydroxyhexanoate) biosynthesis genes of Aeromonas caviae. Journal of

Bacteriology. 1997;**179**(15):4821-4830. Available from: https://pubmed.ncbi.nlm.nih.gov/9244271/

[26] Wen Q, Chen Z, Tian T, Chen W. Effects of phosphorus and nitrogen limitation on PHA production in activated sludge. Journal of Environmental Sciences. 2010;**22**(10):1602-1607

[27] Morgan-Sagastume F, Karlsson A, Johansson P, Pratt S, Boon N, Lant P, et al. Production of polyhydroxyalkanoates in open, mixed cultures from a waste sludge stream containing high levels of soluble organics, nitrogen and phosphorus. Water Research. 2010;**44**(18):5196-5211. Available from: https://pubmed.ncbi.nlm.nih.gov/20638096/

[28] Montiel-Jarillo G, Carrera J, Suárez-Ojeda ME. Enrichment of a mixed microbial culture for polyhydroxyalkanoates production: Effect of pH and N and P concentrations. Science Total Environments. 2017;**583**: 300-307. Available from: https://pubmed.ncbi.nlm.nih.gov/28117150/

[29] Gradíssimo DG, Xavier LP, Santos AV. Cyanobacterial polyhydroxyalkanoates: A sustainable alternative in circular economy. Molecules. 2020;**25**(18):1-23

[30] Antonio GP, Fabbricino PM, Policastro G, Fabbricino ÁM, Fabbricino M, Panico A. Improving biological production of poly (3-hydroxybutyrate-co-3-hydroxyvalerate) (PHBV) co-polymer: a critical review. Review in Environmental Science Bio/Technology. 2021;**20**(2): 479-513. Available from: https://link.springer.com/article/10.1007/s11157-021-09575-z

[31] Sim SJ, Snell KD, Hogan SA, Stubbe JA, Rha C, Sinskey AJ. PHA synthase activity controls the molecular weight and polydispersity of polyhydroxybutyrate in vivo. Nature Biotechnology. 1997;**15**(1):63-67. Available from: https://pubmed.ncbi.nlm.nih.gov/9035108/

[32] Taidi B, Mansfield DA, Anderson AJ. Turnover of poly(3-hydroxybutyrate) (PHB) and its influence on the molecular mass of the polymer accumulated by Alcaligenes eutrophus during batch culture. FEMS Microbiology Letters. 1995;**129**(2–3):201-205. Available from: https://academic.oup.com/femsle/article/129/2-3/201/471642

[33] de Jesus AD, Gomes GVP, da Cunha Pascoal DR, Pinho LS, Chaves LBO, Druzian JI. Simultaneous biosynthesis of Polyhydroxyalkanoates and extracellular polymeric substance (EPS) from crude glycerol from biodiesel production by different bacterial strains. Applied Biochemistry and Biotechnology. 2016; **180**(6):1110-1127. Available from: https://pubmed.ncbi.nlm.nih.gov/27357823/

[34] Fukada E, Ando Y. Piezoelectric properties of poly-β-hydroxybutyrate and copolymers of β-hydroxybutyrate and β-hydroxyvalerate. International Journal of Biological Macromolecules. 1986;**8**(6):361-366

[35] Pötter M, Steinbüchel A. Poly(3-hydroxybutyrate) granule-associated proteins: Impacts on poly(3-hydroxybutyrate) synthesis and degradation. Biomacromolecules. 2005; **6**(2):552-560. Available from: https://pubmed.ncbi.nlm.nih.gov/15762612/

[36] Steinbüchel A, Lütke-Eversloh T. Metabolic engineering and pathway construction for biotechnological production of relevant polyhydroxyalkanoates in

microorganisms. Biochemical Engineering Journal. 2003;**16**(2):81-96

[37] Sudesh K, Abe H, Doi Y. Synthesis, structure and properties of polyhydroxyalkanoates: Biological polyesters. Progress in Polymer Science. 2000;**25**(10):1503-1555. Available from: https://www.researchgate.net/publication/223900137_Synthesis_structure_and_properties_of_polyhydroxyalkanoates_Biological_polyesters

[38] Ishii-Hyakutake M, Mizuno S, Tsuge T. Biosynthesis and characteristics of aromatic polyhydroxyalkanoates. Polymers (Basel). 2018;**10**(11):1267. Available from: www.mdpi.com/journal/polymers

[39] Sharma V, Sehgal R, Gupta R. Polyhydroxyalkanoate (PHA): Properties and modifications. Polymer (Guildf). 2021;**212**:123161

[40] Roja K, Ruben Sudhakar D, Anto S, Mathimani T. Extraction and characterization of polyhydroxyalkanoates from marine green alga and cyanobacteria. Biocatalysis and Agricultural Biotechnology. 2019;**22**:101358. DOI: 10.1016/j.bcab.2019.101358

[41] Carr NG. The occurrence of poly-β-hydroxybutyrate in the blue-green alga, Chlorogloea fritschii. BBA - Biophysical Inclusive Photosynthetic. 1966;**120**(2): 308-310

[42] Panda B, Mallick N. Enhanced poly-β-hydroxybutyrate accumulation in a unicellular cyanobacterium, Synechocystis sp. PCC 6803. Letters Applied Microbiology. 2007;**44**(2): 194-198. Available from: https://pubmed.ncbi.nlm.nih.gov/17257260/

[43] Nishioka M, Nakai K, Miyake M, Asada Y, Taya M. Production of poly-β-hydroxybutyrate by thermophilic cyanobacterium, Synechococcus sp. MA19, under phosphate-limited conditions. Biotechnology Letters. 2001; **23**(14):1095-1099. Available from: https://www.researchgate.net/publication/226809125_Production_of_poly-b-hydroxybutyrate_by_thermophilic_cyanobacterium_Synechococcus_sp_MA19_under_phosphate-limited_conditions

[44] Chen GQ, Zhang G, Park SJ, Lee SY. Industrial scale production of poly (3-hydroxybutyrate-co-3-hydroxyhexanoate). Applied Microbiology and Biotechnology. 2001;**57**(1–2):50-55. Available from: https://pubmed.ncbi.nlm.nih.gov/11693933/

[45] Choi J, Lee SY. Efficient and economical recovery of poly(3-hydroxybutyrate) from recombinant Escherichia coli by simple digestion with chemicals. Biotechnology and Bioengineering. 1999;**62**(5):546-553

[46] Shang L, Jiang M, Chang HN. Poly (3-hydroxybutyrate) synthesis in fed-batch culture of Ralstonia eutropha with phosphate limitation under different glucose concentrations. Biotechnology Letters. 2003;**25**(17):1415-1419. Available from: https://link.springer.com/article/10.1023/A:1025047410699

[47] Kamravamanesh D, Lackner M, Herwig C. Bioprocess engineering aspects of sustainable polyhydroxyalkanoate production in cyanobacteria. Bioengineering. 2018; **5**(4):1-18

[48] Sow S, Ranjan S. Cultivation of Spirulina: An innovative approach to boost up agricultural productivity. ~ 799 ~. Pharma Innovation Journal. 2021;**10**(3): 799-813. Available from: https://doi.org/10.22271/tpi.2021.v10.i3k.5889

[49] Hamed I. The evolution and versatility of microalgal biotechnology: A review. Comprehensive Reviews in Food Science and Food Safety. 2016; **15**(6):1104-1123. Available from: https://onlinelibrary.wiley.com/doi/full/10.1111/1541-4337.12227

[50] Pearce J, Leach CK, Carr NG. The incomplete tricarboxylic acid cycle in the blue-green alga Anabaena variabilis. Journal of General Microbiology. 1969; **55**(3):371-378. Available from: https://pubmed.ncbi.nlm.nih.gov/5783887/

[51] Smith AJ. Modes of cyanobacterial carbon metabolism. Annales de l'Institut Pasteur. Microbiologie. 1983;**134**(1): 93-113. Available from: https://pubmed.ncbi.nlm.nih.gov/6416129/

[52] Steinhauser D, Fernie AR, Araújo WL. Unusual cyanobacterial TCA cycles: Not broken just different. Trends in Plant Science. 2012;**17**(9):503-509. Available from: https://pubmed.ncbi.nlm.nih.gov/22658681/

[53] Xiong W, Brune D, Vermaas WFJ. The γ-aminobutyric acid shunt contributes to closing the tricarboxylic acid cycle in Synechocystis sp. PCC 6803. Molecular Microbiology. 2014;**93**(4):786-796. Available from: https://pubmed.ncbi.nlm.nih.gov/24989231/

[54] Zhang S, Bryant DA. The tricarboxylic acid cycle in cyanobacteria. Science (80-). 2011;**334**(6062): 1551-1553. Available from: https://pubmed.ncbi.nlm.nih.gov/22174252/

[55] Mallick N, Sharma L, Kumar SA. Poly-β-hydroxybutyrate accumulation in Nostoc muscorum: Effects of metabolic inhibitors. Journal of Plant Physiology. 2007;**164**(3):312-317. Available from: https://pubmed.ncbi.nlm.nih.gov/16542751/

[56] Miyake M, Kataoka K, Shirai M, Asada Y. Control of poly-β-hydroxybutyrate synthase mediated by acetyl phosphate in cyanobacteria. Journal of Bacteriology. 1997;**179**(16): 5009-5013. Available from: https://pubmed.ncbi.nlm.nih.gov/9260940/

[57] Morrison SS, Mullineaux CW, Ashby MK. The influence of acetyl phosphate on DspA signalling in the cyanobacterium Synechocystis sp. PCC6803. BMC Microbiology. 2005; **5**(1):1-8. Available from: https://bmcmicrobiol.biomedcentral.com/articles/10.1186/1471-2180-5-47

[58] Juntarajumnong W, Eaton-Rye JJ, Incharoensakdi A. Two-component signal transduction in Synechocystis sp. PCC 6803 under phosphate limitation: Role of acetyl phosphate. Journal of Biochemistry and Molecular Biology. 2007;**40**(5):708-714. Available from: https://pubmed.ncbi.nlm.nih.gov/17927904/

[59] Asada Y, Miyake M, Miyake J, Kurane R, Tokiwa Y. Photosynthetic accumulation of poly-(hydroxybutyrate) by cyanobacteria - the metabolism and potential for CO2 recycling. International Journal of Biological Macromolecules. 1999;**25**(1–3):37-42. Available from: https://pubmed.ncbi.nlm.nih.gov/10416648/

[60] Handrick R, Reinhardt S, Jendrossek D. Mobilization of poly(3-hydroxybutyrate) in Ralstonia eutropha. Journal of Bacteriology. 2000;**182**(20): 5916-5918. Available from: https://pubmed.ncbi.nlm.nih.gov/11004196/

[61] Haase SM, Huchzermeyer B, Rath T. PHB accumulation in Nostoc muscorum under different carbon stress situations. Journal of Applied Phycology. 2012; **2**(24):157-162. Available from: https://www.infona.pl//resource/bwmeta1.

element.springer-9b15f9c1-e4dc-335d-9607-51840513afee

[62] Sharma L, Mallick N. Enhancement of poly-β-hydroxybutyrate accumulation in Nostoc muscorum under mixotrophy, chemoheterotrophy and limitations of gas-exchange. Biotechnology Letters. 2005;**27**(1):59-62. Available from: https://link.springer.com/article/10.1007/s10529-004-6586-1

[63] Sharma L, Mallick N. Accumulation of poly-beta-hydroxybutyrate in Nostoc muscorum: Regulation by pH, light-dark cycles, N and P status and carbon sources. Bioresource Technology. 2005;**96**(11): 1304-1310. Available from: https://pubmed.ncbi.nlm.nih.gov/15734319/

[64] Gao S, Kong Y, Yu J, Miao L, Ji L, Song L, et al. Isolation of axenic cyanobacterium and the promoting effect of associated bacterium on axenic cyanobacterium. BMC Biotechnology. 2020;**20**:61

[65] Mendhulkar VD, Shetye LA. Synthesis of biodegradable polymer polyhydroxyalkanoate (PHA) in cyanobacteria Synechococcus elongates under mixotrophic nitrogen- and phosphate-mediated stress conditions. Indian Biotechnology. 2017;**13**(2):85-93. Available from: https://www.liebertpub.com/doi/10.1089/ind.2016.0021

[66] Costa SS, Miranda AL, Andrade BB, de Assis D, Souza CO, de Morais MG, et al. Influence of nitrogen on growth, biomass composition, production, and properties of polyhydroxyalkanoates (PHAs) by microalgae. International Journal of Biological Macromolecule. 2018;**116**: 552-562. Available from: https://pubmed.ncbi.nlm.nih.gov/29763703/

[67] Kavitha G, Kurinjimalar C, Sivakumar K, Kaarthik M, Aravind R, Palani P, et al. Optimization of polyhydroxybutyrate production utilizing waste water as nutrient source by Botryococcus braunii Kütz using response surface methodology. International Journal of Biological Macromolecules. 2016;**93**(Pt A):534-542. Available from: https://pubmed.ncbi.nlm.nih.gov/27612642/

[68] Kovalcik A, Meixner K, Mihalic M, Zeilinger W, Fritz I, Fuchs W, et al. Characterization of polyhydroxyalkanoates produced by Synechocystis salina from digestate supernatant. International Journal of Biological Macromolecules. 2017;**102**: 497-504 Available from: https://pubmed.ncbi.nlm.nih.gov/28419829/

[69] Panda B, Jain P, Sharma L, Mallick N. Optimization of cultural and nutritional conditions for accumulation of poly-beta-hydroxybutyrate in Synechocystis sp. PCC 6803. Bioresources Technology. 2006;**97**(11): 1296-1301. Available from: https://pubmed.ncbi.nlm.nih.gov/16046119/

[70] Campbell J, Stevens SE, Balkwill DL. Accumulation of poly-beta-hydroxybutyrate in Spirulina platensis. Journal of Bacteriology. 1982;**149**(1):361. Available from: /pmc/articles/PMC216630/?report=abstract

[71] Toh Y, Jau MH, Yew SP, et al. Comparison of polyhydroxyalkanoates biosynthesis, mobilization and the effects on cellular morphology in spirulina platensis and synechocystis sp. uniwg. Journal of Biosciences. 2008; **19**(2):21-38

[72] De Philippis R, Sili C, Vincenzini M. Glycogen and poly-β-hydroxybutyrate synthesis in Spirulina maxima. Journal of General Microbiology. 1992;**138**(8): 1623-1628. Available from: https://www.microbiologyresearch.org/content/journal/micro/10.1099/00221287-138-8-1623

[73] Sudesh K, Taguchi K, Doi Y. Effect of increased PHA synthase activity on polyhydroxyalkanoates biosynthesis in Synechocystis sp. PCC6803. International Journal of Biology Macromolecules. 2002;**30**(2): 97-104. Available from: https://pubmed.ncbi.nlm.nih.gov/11911900/

[74] Chakravarty P, Mhaisalkar V, Chakrabarti T. Study on poly-hydroxyalkanoate (PHA) production in pilot scale continuous mode wastewater treatment system. Bioresources Technology. 2010;**101**(8):2896-2899. Available from: https://pubmed.ncbi.nlm.nih.gov/20045314/

[75] Rahman A, Putman RJ, Inan K, Sal FA, Sathish A, Smith T, et al. Polyhydroxybutyrate production using a wastewater microalgae based media. Algal Research. 2015;**8**:95-98. Available from: https://pennstate.pure.elsevier.com/en/publications/polyhydroxybutyrate-production-using-a-wastewater-microalgae-base

[76] Shrivastav A, Mishra SK, Shethia B, Pancha I, Jain D, Mishra S. Isolation of promising bacterial strains from soil and marine environment for polyhydroxyalkanoates (PHAs) production utilizing Jatropha biodiesel byproduct. International Journal of Biological Macromolecules. 2010;**47**(2): 283-287. Available from: https://pubmed.ncbi.nlm.nih.gov/20417229/

[77] Coelho VC, Klasener Da Silva C, Terra AL, Vieira Costa JA, Greque De Morais M. Polyhydroxy-butyrate production by Spirulina sp. LEB 18 grown under different nutrient concentrations. African Journal of Microbiology Research. 2015;**9**(24):1586-1594. Available from: http://www.academicjournals.org/AJMR

[78] Samantaray S, Mallick N. Impact of various stress conditions on poly-β-hydroxybutyrate (PHB) accumulation in Aulosira fertilissima CCC 444. Current Biotechnology. 2015;**4**(3): 366-372

[79] Samantaray S, Mallick N. Production of poly(3-hydroxybutyrate-co-3-hydroxyvalerate) co-polymer by the diazotrophic cyanobacterium Aulosira fertilissima CCC 444. Journal of Applied Phycology. 2014;**26**(1):237-245

[80] Samantaray S, Mallick N. Production and characterization of poly-β-hydroxybutyrate (PHB) polymer from Aulosira fertilissima. Journal of Applied Phycology. 2012;**24**(4):803-814

[81] Takahashi H, Miyake M, Tokiwa Y, Asada Y. Improved accumulation of poly-3- hydroxybutyrate by a recombinant cyanobacterium. Biotechnology Letters. 2022;**20**(2): 183-186. Available from: https://link.springer.com/article/10.1023/A:1005392827791

[82] Lama L, Nicolaus B, Calandrelli V, Manca MC, Romano I, Gambacorta A. Effect of growth conditions on endo- and exopolymer biosynthesis in Anabaena cylindrica 10 C. Phytochemistry. 1996; **42**(3):655-659

[83] Monshupanee T, Nimdach P, Incharoensakdi A. Two-stage (photoautotrophy and heterotrophy) cultivation enables efficient production of bioplastic poly-3-hydroxybutyrate in auto-sedimenting cyanobacterium. Scientific Reports. 2016;**6**(1):1-9. Available from: https://www.nature.com/articles/srep37121

[84] Madison LL, Huisman GW. Metabolic engineering of poly(3-hydroxyalkanoates): From DNA to plastic. Microbiology and Molecular

Biology Reviews. 1999;**63**(1):21-53. Available from: https://pubmed.ncbi.nlm.nih.gov/10066830/

[85] Khetkorn W, Incharoensakdi A, Lindblad P, Jantaro S. Enhancement of poly-3-hydroxybutyrate production in Synechocystis sp. PCC 6803 by overexpression of its native biosynthetic genes. Bioresources Technology. 2016;**214**:761-768. Available from: https://pubmed.ncbi.nlm.nih.gov/27213577/

[86] Wang B, Xiong W, Yu J, Maness PC, Meldrum DR. Unlocking the photobiological conversion of CO2 to (R)-3-hydroxybutyrate in cyanobacteria. Green Chemistry. 2018;**20**(16): 3772-3782. Available from: https://pubs.rsc.org/en/content/articlehtml/2018/gc/c8gc01208c

[87] Lau NS, Foong CP, Kurihara Y, Sudesh K, Matsui M. RNA-Seq analysis provides insights for understanding photoautotrophic Polyhydroxyalkanoate production in recombinant Synechocystis Sp. PLoS One. 2014;**9**(1): e86368. Available from: https://journals.plos.org/plosone/article?id=10.1371/journal.pone.0086368

[88] Hondo S, Takahashi M, Osanai T, Matsuda M, Hasunuma T, Tazuke A, et al. Genetic engineering and metabolite profiling for overproduction of polyhydroxybutyrate in cyanobacteria. Journal of Bioscience and Bioengineering. 2015;**120**(5):510-517. Available from: https://pubmed.ncbi.nlm.nih.gov/26055446/

[89] Carpine R, Du W, Olivieri G, Pollio A, Hellingwerf KJ, Marzocchella A, et al. Genetic engineering of Synechocystis sp. PCC6803 for poly-β-hydroxybutyrate overproduction. Algal Research. 2017;**25**: 117-127

[90] Suzuki E, Ohkawa H, Moriya K, Matsubara T, Nagaike Y, Iwasaki I, et al. Carbohydrate metabolism in mutants of the cyanobacterium Synechococcus elongatus PCC 7942 defective in glycogen synthesis. Applied and Environmental Microbiology. 2010; **76**(10):3153. Available from: /pmc/articles/PMC2869141/

[91] Zhang S, Qian X, Chang S, Dismukes GC, Bryant DA. Natural and synthetic variants of the tricarboxylic acid cycle in cyanobacteria: Introduction of the GABA Shunt into Synechococcus sp. PCC 7002. Frontiers in Microbiology. 2016;7:1972

[92] Hu Q, Sommerfeld M, Jarvis E, Ghirardi M, Posewitz M, Seibert M, et al. Microalgal triacylglycerols as feedstocks for biofuel production: Perspectives and advances. The Plant Journal. 2008;**54**: 621-639. Available from: https://pubmed.ncbi.nlm.nih.gov/18476868/

[93] Gründel M, Scheunemann R, Lockau W, Zilliges Y. Impaired glycogen synthesis causes metabolic overflow reactions and affects stress responses in the cyanobacterium Synechocystis sp. PCC 6803. Microbiology. 2012;**158**: 3032-3043. Available from: https://pubmed.ncbi.nlm.nih.gov/23038809/

[94] Wu GF, Shen ZY, Wu QY. Modification of carbon partitioning to enhance PHB production in Synechocystis sp. PCC6803. Enzyme and Microbial Technology. 2002. pp. 710-715

[95] Hauf W, Schlebusch M, Hüge J, Kopka J, Hagemann M, Forchhammer K. Metabolic changes in synechocystis PCC6803 upon nitrogen-starvation: Excess NADPH sustains polyhydroxybutyrate accumulation. Metabolites. 2013;**3**(1):101. Available from: /pmc/articles/PMC3901256/

[96] Kourmentza C, Plácido J, Venetsaneas N, Burniol-Figols A, Varrone C, Gavala HN, et al. Recent advances and challenges towards sustainable polyhydroxyalkanoate (PHA) production. Bioengineering. 2017;**4**(2). Available from: www.mdpi.com/journal/bioengineering

[97] Fradinho JC, Domingos JMB, Carvalho G, Oehmen A, Reis MAM. Polyhydroxyalkanoates production by a mixed photosynthetic consortium of bacteria and algae. Bioresources Technology. 2022;**132**:146-153. Available from: https://pubmed.ncbi.nlm.nih.gov/23399498/

[98] Drosg B, Fritz I, Gattermayr F, Silvestrini L. Photo-autotrophic production of poly(hydroxyalkanoates) in cyanobacteria. Chemical and Biochemical Engineering Quarterly. 2015;**29**(2):145-156

[99] Pulz O, Gross W. Valuable products from biotechnology of microalgae. Applied Microbiology and Biotechnology. 2004;**65**(6):635-648. Available from: https://pubmed.ncbi.nlm.nih.gov/15300417/

[100] Gerde JA, Yao L, Lio JY, Wen Z, Wang T. Microalgae flocculation: Impact of flocculant type, algae species and cell concentration. Algal Research. 2014; **3**(1):30-35. Available from: https://www.researchgate.net/publication/259525842_Microalgae_flocculation_Impact_of_flocculant_type_algae_species_and_cell_concentration

[101] Vandamme D, Foubert I, Meesschaert B, Muylaert K. Flocculation of microalgae using cationic starch. Journal of Applied Phycology. 2010; **22**(4):525-530. Available from: https://www.researchgate.net/publication/224946120_Flocculation_of_microalgae_using_cationic_starch

[102] Papazi A, Makridis P, Divanach P. Harvesting Chlorella minutissima using cell coagulants. undefined. 2009;**22**(3): 349-355

[103] Reis MAM, Serafim LS, Lemos PC, Ramos AM, Aguiar FR, Van Loosdrecht MCM. Production of polyhydroxyalkanoates by mixed microbial cultures. Bioprocess and Biosystems Engineering. 2003;**25**(6): 377-385. Available from: https://pubmed.ncbi.nlm.nih.gov/13680343/

[104] Karan H, Funk C, Grabert M, Oey M, Hankamer B. Green bioplastics as part of a circular bioeconomy. Trends in Plant Science. 2019;**24**(3):237-249. Available from: https://pubmed.ncbi.nlm.nih.gov/30612789/

[105] Choi J, Lee SY. Factors affecting the economics of polyhydroxyalkanoate production by bacterial fermentation. Applied Microbiology and Biotechnology. 1999;**51**(1):13-21. Available from: https://link.springer.com/article/10.1007/s002530051357

[106] Lee GN, Na J. Future of microbial polyesters. Microbial Cell Factories. 2013;**12**(1):1-4 Available from: https://microbialcellfactories.biomedcentral.com/articles/10.1186/1475-2859-12-54

[107] Morillo JA, Antizar-Ladislao B, Monteoliva-Sánchez M, Ramos-Cormenzana A, Russell NJ. Bioremediation and biovalorisation of olive-mill wastes. Applied Microbiology and Biotechnology. 2009;**82**(1):25-39

[108] Ansari S, Fatma T. Cyanobacterial polyhydroxybutyrate (PHB): Screening, optimization and characterization. PLoS One. 2016;**11**(6):e0158168. Available from: https://journals.plos.org/plosone/article?id=10.1371/journal.pone.0158168

Chapter 6

Thermochemical Conversion of Algal Based Biorefinery for Biofuel

Arosha Vaniyankandy, Bobita Ray, Subburamu Karthikeyan and Suchitra Rakesh

Abstract

Algae being the photosynthetic organism, currently considered as underexplored species for biofuel production in the entire global region and yet need to be explored more. In presence of algal based theory regarding the thermochemical process, though many researchers have been proceeding with the experiment but have got to stretch it further. This process aims to produce energy and bioactive compounds using algal biomass as a raw material. The current study relates with the thermochemical conversion process and mainly reflects about the algal biomass conversion into biorefinery production, in a short time with easier and economically viable points, unlike other biochemical and chemical conversion processes. In thermochemical process, high temperatures used during the process produces different biofuels including solid, liquid, gaseous biofuels. This thermal decomposition process of algal biomass can be categorized into Gasification, Pyrolysis, Direct combustion, Hydrothermal process, and Torrefaction. Hence, in this study, it briefs on different type of processes for better production of biofuel as well as its significant merit and demerit comparisons of each process.

Keywords: algae, biomass, thermochemical conversion, biorefinery, liquefaction

1. Introduction

Algae, grouped among the photosynthetic organism, are sustained in the diverse form of habitats. It can flourish in freshwater, marine water as well as wastewater. Algae are a suitable biomass resource for renewable energy production because of the rapid growth rate, high content of lipids, and tremendous biomass productivity [1]. The algal biorefinery concept integrates various processes for converting algal biomass into biofuels and other bioactive products [2]. They aim to produce energy and bioactive compounds using algal biomass as a raw material. The conversion process for biorefinery production includes biochemical, chemical, and thermochemical conversion processes.

The thermochemical conversion process is considered an efficient method for producing biofuel from algal biomass. During the process, molecules in algal biomass are broken down to release their potential energy. It transforms the entire algal biomass to the respective fuel in a shorter time, unlike other conversion processes. The process

uses high temperatures to degrade the algal biomass to produce different biofuels, including solid, liquid and gaseous biofuels [3]. It is the best option to process algae with low lipid content or residues after extraction of the algae with high lipid content. The process is direct, easy, and fast compared to biochemical and chemical conversion processes and is economically viable [4]. This thermal decomposition process of algal biomass can be categorized into Gasification, Pyrolysis, Direct combustion, Hydrothermal process, and Torrefaction.

Gasification is an excellent process to convert algal biomass to gaseous fuels. In contrast, pyrolysis and hydrothermal liquefaction (HTL) processes give bio-oil low molecular weight and bio-crude high energy density [5]. In the gasification process, the partial oxidation of algal biomass occurs at high temperatures. Combustible fuel gases like CH_4 and H_2 are produced. The actual process of gasification involves the reaction with the carbon-containing compound in the algal biomass and the air, oxygen or steam present at high temperatures in the gasifier, resulting in the production of syngas and mixtures of other combustible gases like CO, H_2, CO_2, N, CH_4 [2]. These studies indicated the promising role of steam gasification in hydrogen production. Pyrolysis is an anaerobic heating process that produces medium to low calorific value liquid fuels on a large scale. The significant products obtained after pyrolysis are bio-oil, biochar, and charcoal [4, 6].

Hydrothermal processes such as liquefaction is emerged to be the most promising method to convert wet algal biomass to liquid fuel with the use of high temperature and pressure. It consists of evolving technique that can connect biomass with high moisture content and low energy and can convert into heat, hydrogen, biochar, electricity and other type of synthetic fuels. It is more efficient and favorable in converting wet algal biomass to biofuel than pyrolysis [7]. Combustion is the easiest and most traditional method among all thermochemical processes. The direct combustion process involves burning or incinerating the algal biomass and converting the stored chemical energy in the biomass into gases in the presence of excess air [4, 8]. Whereas pyrolysis and combustion characteristics of *Chlorella vulgaris* are under different heating rates found compared to pyrolysis, combustion produces higher biomass, and the faster heating rate leads to the quicker and higher conversion [9]. The torrefaction process is introduced to overcome the demerit of low calorific values of algae. These upgrading methods involve the thermal degradation of algal Sbiomass in an inert or N_2 environment [2].

Biofuel generated from algae will be environmentally friendly, non-toxic, and highly biodegradable. So these are considered a better alternative to fossil fuel as it has many disadvantages like environmental degradation, climate change, rising price, and depletion. The algal biorefinery approach is an excellent way to produce biofuels and other value-added products from algae. Many review papers reviewed different processes and steps involved in algae-based biorefinery. Since solid, liquid, and gaseous fuels can be produced via thermochemical strategies, these are emerged as the viable option to recover energy from algal biomass. The thermochemical conversion process can recover highly efficient and economically valuable biofuels. The thermochemical conversion process provides a simpler route of conversion. Various thermochemical approaches are widely explored because of their huge advantage over other methods. This chapter describes the different types of thermochemical conversion process for various biorefinery productions as well as it also emphasizes the influence of catalysts in thermochemical process for upgrading of biofuels. For a brief understanding of this chapter a figure have been shown below mentioned as **Figure 1**.

Figure 1.
Algal biomass to biofuel conversion techniques.

2. Algal-based biorefinery

Algal-based biorefinery is a cost-effective approach to producing biofuels, bioenergy, and other value-added products by integrating algal biomass conversion processes and equipment [10]. It adds to the concept of converting algal biomass into useful, commercially important products and energy. The major stages in algal biorefinery include upstream and downstream processing, such as cultivation, harvesting, drying, and conversion processes to produce biofuel and other value-added products.

Algal cultivation becomes economically feasible since algae can be grown in wastewater as a culture medium to cultivate algae. The importance and necessity of aquaculture wastewater for the purpose of cultivating algae and even highly flourished growth of microalgae in fertilizer wastewater leads to the production of biodiesel from algal biomass in a cost-effective way [11, 12]. Open raceway ponds and closed photo-bioreactors comprise the principal method for algal cultivation. Compared with other algal culture systems, open culture systems are cost-effective and easy to install and maintain, and their energy consumption is preferably lower. The negative impact of this system is a lack of control over water temperature, light intensity, and evaporation [13, 14]. Whereas in the case of a closed culture system, photobioreactors can produce 3–5 times more biomass. It can cultivate single species of microalgae in a considerable quantity. Tubular, flat plate, column, and membrane photobioreactors are different types of closed systems [14]. A novel, cost-effective algal cultivation strategy, mixotrophic microalgae biofilm, was introduced to improve productivity [15].

The size of algae is relatively minute in particular, and its negative surface charge makes the separation process difficult, making it challenging for harvesting. Several techniques are applied to neutralize these negative charges [16]. Algae harvesting from

the aqueous suspension can be done mechanically, chemically, biologically, or using electrical-based methods. A combination of two or more of these methods is also used [17]. Different technologies are used to harvest algal biomass, including centrifugation, flocculation, bio-flocculation, flotation, filtration, gravity sedimentation and electrocoagulation. Another cost effective method of easiest harvesting is combination of flocculation-sedimentation cum centrifugation [15, 16]. Partial harvesting of algal biomass with vacuum gas lift prior to the complete harvesting (centrifugation) proved efficient and cost-effective [18–20]. In another harvesting experiment, auto flocculation uses appropriate flocculants like poly aluminum chloride, aluminum sulfate, and pH adjusted chitosan is the best and economical way to harvest the microalgae. Harvesting efficiency can also be enhanced by adding auto flocculating microalgae, which can induce faster sedimentation of non-flocculating microalgae [21].

Drying can be done to protect the algal biomass from spoilage. For the hydrothermal process, the algal biomass need not be dried because the process is carried out in the water and requires 95% moisture content. The other thermochemical processes like pyrolysis, gasification, and combustion needs to be dried algal biomass to produce biofuel and high value products [17]. The significant algae drying process comprises rotary dryer, solar heat drying, spray drying, cross flow, and vacuum shelf drying [22]. Among that, solar heat drying or sun drying is the most basic drying with a low cost of budget but requires more duration time to dry. Algal biomass is disrupted in order to release intracellular biomolecules. Nowadays, mechanical and non-mechanical cell disruption methods are used to disrupt the algal cell wall. Non-mechanical methods comprise a chemical method, osmotic shock, and treatment using enzymes and detergents. Osmotic shock involves applying a high concentration of a solute, such as a dextran, salts, or polyethylene glycol, around a cell to lower its osmotic pressure. These cause disruption of the algal cell wall and the release of intracellular molecules. Moreover, hypotonic osmotic shock can damage the membrane of all algal species but not the cell wall [23]. Chaetoceros mueller algae produced 35% methane and 72% algal lipid in an osmotic shock experiment [24]. Cell disruption can also occur using various chemicals such as organic solvents, surfactants, hypochlorite, and chelating agents. Acids and alkali treatments are also used for the algal cell disruption. Several parameters were studied and optimized in order to increase lipid extraction potency from Scenedesmus sp. (cellulase, pectinase, xylanase, protein concentration, pH, temperature, and incubation time) [25]. In the case of the enzymatic cell disruption method, enzymes are used to recover intracellular components. It can degrade cell wall components such as cellulose, hemicellulose, alginates, and glycoproteins. Mechanical methods in the form of liquid and solid shearing (bead milling, high-speed homogenizer, and high-pressure homogenizer), energy transfer (ultrasonication, microwave, and laser), and heat (thermolysis and autoclaving) and as a current (pulsed electric field) are considered as an alternative method to disrupt the cell wall of algae [26]. The bead milling method induces direct mechanical damage to the algal cell. These cells are damaged by applying forces from collisions between cells and beads. The collision is propped up with the help of a rotating shaft in the grinding chamber [27]. Another technique method is ultrasonication which uses ultrasound waves to disrupt algal cells.

Similarly, the pulsed electric field technique uses an external electric field, creating a critical electric potential across the algal cell wall, thereby causing disruption of the cell wall. Heat treatment methods such as autoclaving and thermolysis are also effective for cell disruption [28]. Many valuable biomolecules can be extracted from

algae by cell disruption methods. After the cell disruption process, the extraction process begins. Supercritical fluids and deep eutectic solvents are used in solvent extraction. The organic solvent extraction technique is a well-known method for the extraction of algal biomolecules. This technique enhances the extraction yield by facilitating the access of solvents to inner cellular molecules. In addition to terpenes, liquid polymers, ionic liquids, and deep eutectic solvents, bio-based solvents are used for solvent extraction. In the case food and pharmaceutical industries supercritical fluid extraction technique is primarily employed as it is a contamination-free method of extraction. Separation and purification methods are done to separate the impurities and the molecules of least interest. Separation methods to purify the extracted components include electrophoresis, membrane separation, ultracentrifugation, etc. [26].

Various conversion technologies are employed to convert algal biomass into value-added products, including biochemical, chemical, and thermochemical technologies. Biochemical conversion of algal biomass is achieved through biological treatments to produce biofuels. These conversion methods include fermentation, anaerobic digestion, and transesterification. Anaerobic digestion converts algal biomass to hydrogen and methane, while fermentation produces ethanol, acetone, and butanol; transesterification produces biodiesel.

3. Various processes of thermochemical conversion from algal biomass

The thermochemical conversion process is known to be an efficient method for the conversion of algal biomass into biofuel. It involves the thermal degradation of the biomass structure. From the evidence of many journals, chemical and biochemical methods are utilized in conversion to biofuel, whereas these days thermochemical conversion is also commonly used as it provides a more straightforward route to synthesize biofuel. The following thermochemical conversion processes are into Gasification, Pyrolysis, Direct combustion, Hydrothermal process, and Torrefaction. These processes also consist of demerits followed by merit points where the differences are shown in **Table 1**.

3.1 Gasification

As mentioned earlier, the gasification process is the partial oxidation of algal biomass that prefers to work only at high temperatures along with the combustible fuel. The syngas, basically produced by the gasification process, has a low calorific value of 4–6 MJ/m^3 and can be used as a fuel for gas engines or gas turbines. The gasification process also produces hydrocarbon compounds which can be further converted into methanol *via*. The Fischer Tropsch conversion pathway. To effectively perform the gasification process, the moisture content of the biomass should be less than 14% [3]. In a study, it had pointed out that 40% of the moisture content in the algal biomass can be tolerated by the gasifier through the comparative performance analysis. It was also shown that this moisture content of the biomass is considered to be an important factor influencing the heating value of gas and even the high moisture content seriously affecting the performance of the gasifier. At 5% moisture content, the high heating value and the cold gas efficiency of the syngas produced are 5.138 MJ/kg and 73.81%. At 30% moisture content, it would be 3.338 MJ/kg and 44.24% [29].

Thermochemical Process	Merit	Demerit
Gasification	1. Converts algal biomass into gaseous fuels. 2. Combustible fuel gas like CH_4 and H_2 are produced. 3. Syngas produced can be used as a fuel for gas engines or gas turbines.	1. Moisture content of biomass should be less than 14%. 2. High moisture content in biomass affects the performance. 3. Low large scale production.
Pyrolysis	1. Anaerobic heating process that give rise to bio-oil having low molecular weight and bio crude having high energy density. 2. Produces medium to low calorific value liquid fuels in large scale. 3. Bio-oil, biochar, and charcoal are obtained even in moderate temperature range from 350 to 700°C.	1. Expensive process. 2. Requires high energy and temperature for conversion. 3. Slow in process.
Direct Combustion	1. Easiest and traditional method. 2. Involves burning or incineration of biomass. 3. Converts stored chemical properties present in biomass into gas state.	1. Requires high temperature, capacity to carry out is 800°C. 2. Requires pretreatment process like chopping, grinding, drying. 3. Basically leads to more energy and high cost.
Hydrothermal Process	1. Converts wet algal biomass into liquid fuel. Water volatile favorable. 2. Hydrothermal carbonization requires mild temperature and pressure to produce biochar. 3. This process can be carried out in low temperature, *i.e.,* 300–350°C.	1. Expensive in process. 2. Forms corrosion. 3. Forming of tar and coke.
Torrefaction	1. Designed to improve drawback of algal biomass poor calorific value. 2. Also improves the physiochemical properties as well as fuel characteristics of algal biomass. 3. Also referred to as mild pyrolysis that give rise to solid biochar.	1. Low amount of density enhancement is applicable. 2. Applied energy density during the process is not improved. 3. Water volatile not favorable.

Table 1.
Comparison among various thermochemical processes.

3.2 Pyrolysis

Through the process of pyrolysis, the algal biomass is converted into bio-oil, syngas, and charcoal in the absence of air. It is an anaerobic heating process, and heating can be done at a moderate temperature range of 350–700°C. The pyrolysis can be categorized into fast, flash, and slow pyrolysis on the basis of operating conditions. The production of bio-oil and biochar can be achieved by performing fast pyrolysis, and slow pyrolysis results in the production of pyrolytic gas and biochar. Slow pyrolysis having a heating rate range of 0.1 and 1°C/S with the sample particle size ranging between 5 and 50 mm,

allows the production of solid, liquid and gaseous products. Fast pyrolysis gives rise to liquid and gaseous products. Since having a high heating rate, flash pyrolysis gives liquid products [30]. Microwave-induced pyrolysis is carried out from the microalga *Scenedesmus almeriensis* in an electric furnace and showed that the microwave-induced pyrolysis gives rise to higher syngas and H_2 production [22, 23].

3.3 Direct combustion

The combustion process is said to be the easiest among all thermochemical processes. Both microalgae and macroalgae residues while heating follows into lipid extraction which is termed as an effective method [31–33]. Combustion is usually carried out at a temperature. But the capacity to carry out a temperature is around 800°C in the boiler, that furnaces or steam turbines and used to generate electricity. The major products generated after the combustion processes include CO_2, H_2O, and heat. The major disadvantage of this process involves that it requires pretreatment processes like chopping, drying, and grinding, which utilizes more energy and leads to high cost. Also the presence of impurities in biomass such as sodium, potassium, sulfur and nitrogen leads to problems with fouling and corrosion [34].

Various studies have been done in the combustion of microalgae. Among the study used *Haematococcus pluvialis* microalgae (M) and the chemical extraction residue (MR). A couple of TG-MS systems were used to investigate the combustion and emission properties of M and MR and the results revealed that the combustion of M and MR took place in three stages i.e. the decomposition of proteins, carbohydrates, lipids, and char was the first stage, followed by the volatilization of free water and a tiny amount of volatiles, and finally the decomposition of minerals. Whereas co-combustion of *C. vulgaris*, industrial waste of textile dyeing sludge (TDS) and their blends were also included in few of the studies [24, 26].

3.4 Hydrothermal process

HTL is emerged to be the most promising method to convert wet algal biomass to liquid fuel and various value-added products. The process is carried out at a low temperature, usually 300–350°C, and high pressure (5-20 MPa) condition with the help of a catalyst and in the presence of hydrogen and yields bio-oil [35, 36]. The process effectively converts the biomass with water activity into smaller molecular components with high energy densities. The drawback of the conventional HTL method paves the way for the two-stage sequential hydrothermal liquefaction (SEQHTL) method, which overcome the limitation of the conventional method in recovering bioactive compounds [37].

In an experiment given as an example, nine species of algae were selected in order to perform HTL at temperatures of 280°C and 320°C to find out the effect of the biochemical composition of the species on bio-oil yields and properties at two different temperatures. They got maximum bio-oil yield at a temperature of 320°C in the algae *Nannochloropsis,* which contains high lipid content [38]. It has been found through a microchip known to control high temperature and pressure that allows the HTL process in situ using fluorescence microscopy [39]. It requires a thermochemical process to convert the algal biomass into biochar products. The process involves heating algal biomass in water at the temperature of 200°C under pressure less than 2Mpa within 60 min of residence time. The process is exothermic and spontaneous [40]. In an experiment, lipid was extracted from Picochlorum oculatum. It was used as an algal

biomass for the conversion of algal hydrochar via hydrothermal carbonization and the resultant hydrochar were found to be a promising adsorbent for metal remediation of wastewater [41].

3.5 Torrefaction

The torrefaction process is designed to offset the drawback of microalgae's poor calorific value. These are the pretreatment process to improve the physicochemical properties of algal biomass and thereby improve the fuel characteristics of algal biomass. The process involves the thermal degradation of algal biomass in an inert or nitrogen environment at one atmospheric pressure and 200–300°C temperature at a residence time of 10 to 60 min [42]. The torrefaction process gives rise to solid biochar. The efficiency of the process can be influenced by certain factors such as temperature, residence time, and composition of the biomass [43]. The torrefaction process shows high similarity with pyrolysis, but the process needs low operating temperatures, so it is called mild pyrolysis [44]. During the process, carbohydrates, proteins, and lipids are all degraded at varying rates resulting in partial carbonization. Few algae such as *Chlorella sp., Nanochloropsis sp.* are analyzed and their thermal degradation of carbohydrates, proteins, and lipids are demonstrated where the activation energies of carbohydrates, lipids, and proteins are in the range of 53.28–53.30, 142.61–188.35 and 40.21–59.23 KJ/mol and the thermal degradation of carbohydrates, proteins, and lipids, are in temperature ranges of 164–497, 209–309, and 200–635°C, respectively. Torrefaction is classified into conventional, microwave, wet, and oxidative torrefaction. These are again categorized as light (200–235°C), mild (235–275°C), and severe (275–300°C) torrefaction depending on the torrefaction temperatures [36–38].

4. Upgrading of bio-oil in pyrolysis and hydrothermal liquefaction

The bio-oil obtained from the HTL and pyrolysis process is considered a best-suited alternative to petroleum if and only if the quality of the bio-oil is enhanced. The bio-oil extracted after the thermochemical conversion process contains phenols, acids, aldehydes, N, and O heteroatoms which confer thermal stability and corrosion. The use of bio-oils is restricted due to the high oxygen content, strong acidity, and high calorific value of bio-oil. Due to these reasons, the up-gradation of bio-oil is essential, which involves enhancing the quality of bio-oil to use in transportation.

4.1 Emulsification

The simple upgrading method involves the emulsification of bio-oil with other fuels. However, bio-oil is immiscible with petroleum-based fuels and can be emulsified with biodiesel using surfactants. As a liquid fuel, upgrading bio-oil by emulsifying it with diesel oil reduces viscosity and enhances the calorific value and cetane number [45, 46].

Therefore, the use of a cheap and appropriate emulsifying agent is essential in bio-oil upgrading through emulsification. A study said emulsions of bio-oil with biodiesel and showed that the production of the most stable emulsion was acquired using the surfactant class polyethylene glycol-di-polyhydroxy stearate (PEG-DPHS), having an HLB number of 4.75 and a mass ratio of 32:8:1 diesel: bio-oil: surfactant. Even while using the co-surfactant SPAN80 in addition to the surfactant showed that the ability to solubilize bio-oil in diesel increases with increasing cosurfactant/surfactant ratio

[46]. When compared to the original bio-oil, in case of diesel emulsions possessed more fuel properties. These are very simple and rapid upgrading methods but expensive due to the addition of surfactant and high energy costs.

4.2 Esterification

Esterification or otherwise called alcoholysis, is the process of conversion of free fatty acids into their respective alkyl esters. The bio-oil produced contains organic acids, which contributes to acidity, instability, and a high degree of unsaturation and can be reduced by the process of esterification. The reaction between the fatty acids and alcohol at atmospheric pressure with the help of catalysts gives rise to the formation of alkyl ester or biodiesel. Bio-oil also consists of aldehydes possessing challenges for bio-oil upgrading through esterification [40, 41]. In some study, ozone oxidation technology is used to pretreat bio-oil for the conversion of aldehydes into acids. And another through the experiment demonstration the two-step esterification-hydrogenation process showed better performance in bio-oil upgrading than the one-step esterification-hydrogenation process, and it provides higher alcohol and more stable compounds [42, 43].

4.3 Hydrogenation

Bio-oil derived after the thermochemical conversion process contains high oxygen content, which can be removed using high-pressure hydrogen, known as hydrogenation. The hydrogenation reaction is carried out during hydrotreating which increases the hydrogen content, thereby increasing the quality of bio-oil. Hydrotreating is a refinery process that aims to reduce bio-oil's N, O, and S contents. Using a catalytic process with high-pressure hydrogen, it eliminates oxygen as water. In a similar way, high consistency of pure nitrogen enhances to form ammonia synthesis. The energy and heat basically utilized here are recirculated easily and recovers it for power generation [44, 47]. Whereas in another case, two-step esterification hydrogenation even helps in upgrading the bio-oil. It basically helps to degrade the active compounds mostly acids and ketones and rather helps in raising the contents of alcohols and esters [48].

4.4 Cracking

These are the upgrading process to convert the oxygen content in the bio-oil to H_2O, CO, and CO_2 using catalysts. The reaction occurs in a fixed or fluidized bed reactor system under normal pressure. Zeolite catalyst (HZSM-5) is the most common catalyst used in catalytic cracking due to its strong acidity, high reactivity, and stable porous structure [40, 49]. An experiment had proposed that the bio-oil upgradation can be done with two heating units with or without the presence of zeolite catalyst but the characteristics of catalytic cracked bio-oil were better than the non-catalytic cracked bio-oil [47]. During the bio-oil upgrading, the formation of coke can deactivate catalysts and its significant issue. Another experiment conducted with catalytic cracking of bio-oil models such as acetic acid, cyclopentanone and guaiacol had been investigated for the formation of coke using fixed bed reactor. It has found that compared to cyclopentanone and acetic acid, guaiacol produces more coke as it has ring structures that directs polymerization on the catalyst surface to form coke [50]. In **Table 2**, the thermochemical process with its various supported catalyst for the production of biorefinery products have been shown already.

Algal Species Name	Thermochemical process utilized	Temperature inhibited	Catalyst used	Biorefinery products	Other beneficial substances	Reference
Chlorella vulgaris	Super Critical Water Gasification	600°C	NA	NA	P accumulation; Organic compounds decomposition	[51]
Saccharina japonica	Gasification	<500°C	Alkaline Thermal Treatment (ATT); Ni/ZrO_2	H_2	NA	[52]
Fucus serratus, Laminaria digitate and Nannochloropsis oculata	Steam Gasification and Pyrolysis	800°C	Fe_2O_3-CeO_2 > Red mud > Activated Red Mud	H_2	Tar degradation	[6]
Scenedesmus sp. and *Spirulina sp.*	Hydrothermal Liquefaction and Slow Pyrolysis	300°C and 450°C	NA	Bio-oil	NA	[8]
Fucus vesiculosus	Hydrothermal Liquefaction	300°C	Hβ zeolite	Biocrude-oil	NA	[53]
C. vulgaris	Hydrothermal Liquefaction	350°C	NA	Biocrude	NA	[48]
Nannochloropsis oceanica	Torrefaction	300°C	Potassium carbonate	Biofuel	NA	[54]
C. vulgaris	Torrefaction	300°C	NA	Biochar	Methylene blue adsorption	[55]
Ascophyllum nodosum	Hydrothermal carbonization	300°C	$ZnCl_2$	Biochar	Antibiotic removal from water	[56]
Arthrospira platensis	Fast Pyrolysis	800°C	Zeolites	Biofuel	Benzene, toluene, xylene, cyclobutane, acetonitrile	[50]

Table 2.
Effect of various thermochemical process on microalgae.

4.5 Steam reforming

These are the promising method to produce hydrogen and syngas from algal biomass. The bio oil is kept in steam at high temperature. In the steam reforming process, fluidized or fixed bed reactor system is always used at the temperature of 700–1000°C using catalyst. Nickel is used widely as a catalyst for steam reforming [57].

5. Conclusion

Algae being the main source of feedstock for the biorefinery production have helped not only in maintaining sustainability but also keeping it pollution free. Since algae considered as third generation for the production of biofuel, until now many researches have shown evidences with many positive effective work that helped both in human as well as in living environment. Thermochemical conversion process found to be a promising route as it can connect with algal based biorefinery production. Basically it consists of recovering energy for conversion from algal biomass. Mostly thermochemical processes such as gasification, combustion, pyrolysis works based on less moisture content samples but hydrothermal process compared to other processes can be proceed with wet algal biomass (high moisture content). In some of the processes, catalyst containing of chemical or biochemical are added for better result of biorefinery production in order to upgrade the bio-oil formed during the thermochemical process. Hence, though this process consists some positive effect but it also has its negative impact too where in some processes both wet and dry has its own impact along with large scale production issues. But overall, these thermochemical merit and demerit process leads to great study for research for future bio-refinery production.

Acknowledgements

Towards Biofuel Research Laboratory of Department of Microbiology, Central University of Tamil Nadu, India and to SERB DST Project.

Author details

Arosha Vaniyankandy[1], Bobita Ray[1], Subburamu Karthikeyan[2] and Suchitra Rakesh[1*]

1 Biofuel Research Laboratory, Department of Microbiology, School of Life Sciences, Central University of Tamil Nadu, Thiruvarur, Tamil Nadu, India

2 Biomass Refining Laboratory, Department of Renewable Energy Engineering, Tamil Nadu Agricultural University, Coimbatore, Tamil Nadu, India

*Address all correspondence to: suchitrar@cutn.ac.in

References

[1] Alam F, Mobin S, Chowdhury H. Third generation biofuel from algae. Procedia Engineering. 2015;**105**:763-768

[2] Chandra R, Iqbal HMN, Vishal G, et al. Algal biorefinery: A sustainable approach to valorize algal-based biomass towards multiple product recovery. Bioresource Technology. 2019;**278**:346-359

[3] Chen WH, Lin BJ, Huang MY, et al. Thermochemical conversion of microalgal biomass into biofuels: A review. Bioresource Technology. 2015;**184**:314-327

[4] Koyande AK, Show P-L, Guo R, et al. Bio-processing of algal bio-refinery: A review on current advances and future perspectives. Bioengineered. 2019;**10**(1):574-592. DOI: 10.1080/21655979.2019.1679697

[5] Fan L, Zhang H, Li J, et al. Algal biorefinery to value-added products by using combined processes based on thermochemical conversion: A review. Algal Research. 2020;**47**:101819

[6] Duman G, Uddin MA, Yanik J. Hydrogen production from algal biomass via steam gasification. Bioresource Technology. 2014;**166**:24-30. DOI: 10.1016/j.biortech.2014.04.096

[7] Rout PR, Goel M, Mohanty A, et al. Recent advancements in microalgal mediated valorisation of wastewater from hydrothermal liquefaction of biomass. Bioenergy Research. 2022:1-16. DOI: 10.1007/s12155-022-10421-5

[8] Vardon DR, Sharma BK, Blazina GV, et al. Thermochemical conversion of raw and defatted algal biomass via hydrothermal liquefaction and slow pyrolysis. Bioresource Technology. 2012;**109**:178-187

[9] Agrawal A, Chakraborty S. A kinetic study of pyrolysis and combustion of microalgae Chlorella vulgaris using thermo-gravimetric analysis. Bioresource Technology. 2013;**128**:72-80

[10] Ghatak HR. Biorefineries from the perspective of sustainability: Feedstocks, products, and processes. Renewable and Sustainable Energy Reviews. 2011;**15**:4042-4052

[11] Ray B, Rakesh S. Phycoremediation of aquaculture wastewater and algal lipid extraction for fuel conversion. Highlights in BioScience. 2022:5

[12] Ambat I, Bec S, Peltomaa E, et al. A Synergic Approach for Nutrient Recovery and Biodiesel Production by the Cultivation of Microalga Species in the Fertilizer Plant Wastewater. Scientific Reports. 2019;**9**(1):1-9

[13] Jankowska E, Zieliński M, Dębowski M, et al. Anaerobic digestion of microalgae for biomethane production. In: Second and Third Generation of Feedstocks: The Evolution of Biofuels. Elsevier. 2019:405-436

[14] Siddiki SYA, Mofijur M, Kumar PS, et al. Microalgae biomass as a sustainable source for biofuel, biochemical and biobased value-added products: An integrated biorefinery concept. Fuel. 2022;**307**:121782. DOI: 10.1016/j.fuel.2021.121782

[15] Roostaei J, Zhang Y, Gopalakrishnan K, et al. Mixotrophic microalgae biofilm: A novel algae cultivation strategy for improved productivity and cost-efficiency of

biofuel feedstock production. Scientific Reports. 2018;**8**:1-10. DOI: 10.1038/s41598-018-31016-1

[16] Malik S, Shahid A, Haider MN, et al. Prospects of multiproduct algal biorefineries involving cascading processing of the biomass employing a zero-waste approach. Current Pollution Reports. 2022;**8**:2. [cited July 4, 2022], 147-158. DOI: 10.1007/s40726-022-00213-y

[17] Aliyu A, Lee JGM, Harvey AP. Microalgae for biofuels via thermochemical conversion processes: A review of cultivation, harvesting and drying processes, and the associated opportunities for integrated production. Bioresource Technology Reports. 2021;**14**:100676

[18] Barros AI, Gonçalves AL, Simões M, et al. Harvesting techniques applied to microalgae: A review. Renewable and Sustainable Energy Reviews. 2015;**41**:1489-1500

[19] Tan XB, Lam MK, Uemura Y, et al. Cultivation of microalgae for biodiesel production: A review on upstream and downstream processing. Chinese Journal of Chemical Engineering. 2018;**26**:17-30

[20] Barrut B, Blancheton JP, Muller-Feuga A, et al. Separation efficiency of a vacuum gas lift for microalgae harvesting. Bioresource Technology. 2013;**128**:235-240

[21] Şirin S, Trobajo R, Ibanez C, et al. Harvesting the microalgae Phaeodactylum tricornutum with polyaluminum chloride, aluminium sulphate, chitosan and alkalinity-induced flocculation. Journal of Applied Phycology. 2012;**24**:1067-1080

[22] Show KY, Lee DJ, Chang JS. Algal biomass dehydration. Bioresource Technology. 2013;**135**:720-729

[23] Halim R, Papachristou I, Kubisch C, et al. Hypotonic osmotic shock treatment to enhance lipid and protein recoveries from concentrated saltwater Nannochloropsis slurries. Fuel. 2021;**287**:119442

[24] González-González LM, Astals S, Pratt S, et al. Osmotic shock pre-treatment of Chaetoceros muelleri wet biomass enhanced solvent-free lipid extraction and biogas production. Algal Research. 2021;**54**:102177

[25] Zhang Y, Kong X, Wang Z, et al. Optimization of enzymatic hydrolysis for effective lipid extraction from microalgae Scenedesmus sp. Renewable Energy. 2018;**125**:1049-1057

[26] Corrêa PS, Morais Júnior WG, Martins AA, et al. Microalgae biomolecules: Extraction, separation and purification methods. Processes. 2020;**9**(1):10

[27] Wang M, Chen S, Zhou W, et al. Algal cell lysis by bacteria: A review and comparison to conventional methods. Algal Research. 2020;**46**:101794

[28] Timira V, Meki K, Li Z, et al. A comprehensive review on the application of novel disruption techniques for proteins release from microalgae. Critical Reviews in Food Science and Nutrition. 2022;**62**(16):4309-4325

[29] Ramzan N, Ashraf A, Naveed S, et al. Simulation of hybrid biomass gasification using Aspen plus: A comparative performance analysis for food, municipal solid and poultry waste. Biomass and Bioenergy. 2011;**35**:3962-3969

[30] Marcilla A, Catalá L, García-Quesada JC, et al. A review of thermochemical conversion of microalgae. Renewable and Sustainable Energy Reviews. 2013;**27**:11-19

[31] Beneroso D, Bermúdez JM, Arenillas A, et al. Microwave pyrolysis of microalgae for high syngas production. Bioresource Technology. 2013;**144**:240-246

[32] Beneroso D, Monti T, Kostas ET, et al. Microwave pyrolysis of biomass for bio-oil production: Scalable processing concepts. Chemical Engineering Journal. 2017;**316**:481-498

[33] Fang P, Gong Z, Wang Z, et al. Study on combustion and emission characteristics of microalgae and its extraction residue with TG-MS. Renewable Energy. 2019;**140**:884-894. DOI: 10.1016/j.renene.2019.03.114

[34] Gai C, Liu Z, Han G, et al. Bioresource technology combustion behavior and kinetics of low-lipid microalgae via thermogravimetric analysis. Bioresource Technology. 2015;**181**:148-154. DOI: 10.1016/j.biortech.2015.01.045

[35] Peng X, Ma X, Xu Z. Thermogravimetric analysis of co-combustion between microalgae and textile dyeing sludge. Bioresource Technology. 2015;**180**:288-295

[36] Guo B, Yang B, Silve A, et al. Hydrothermal liquefaction of residual microalgae biomass after pulsed electric field-assisted valuables extraction. Algal Research. 2019;**43**:101650

[37] Martinez-Fernandez JS, Chen S. Sequential hydrothermal liquefaction characterization and nutrient recovery assessment. Algal Research. 2017;**25**:274-284

[38] Shakya R, Adhikari S, Mahadevan R, et al. Influence of biochemical composition during hydrothermal liquefaction of algae on product yields and fuel properties. Bioresource Technology. 2017;**243**:1112-1120

[39] Cheng X, Ooms MD, Sinton D. Biomass-to-biocrude on a chip via hydrothermal liquefaction of algae. Lab on a Chip. 2016;**16**:256-260. DOI: 10.1039/C5LC01369K

[40] Heilmann SM, Davis HT, Jader LR, et al. Hydrothermal carbonization of microalgae. Biomass and Bioenergy. 2010;**34**:875-882

[41] Ibtisam Abbasi. Conversion of Residual Algal Biomass into Hydrochar. Azo Materials. 2022. Available from: https://www.azom.com

[42] Chiaramonti D, Prussi M, Buffi M, et al. Thermochemical conversion of microalgae: Challenges and opportunities. Energy Procedia. 2015;**75**:819-826

[43] Shankar Tumuluru J, Ghiasi B, Soelberg NR, et al. Biomass Torrefaction process, product properties, reactor types, and Moving Bed Reactor Design Concepts. Frontiers in Energy Research. 2021;**9**:728140

[44] Chen WH, Kuo PC. A study on torrefaction of various biomass materials and its impact on lignocellulosic structure simulated by a thermogravimetry. Energy. 2010;**35**:2580-2586

[45] Ho SH, Zhang C, Tao F, et al. Microalgal Torrefaction for solid biofuel production. Trends in Biotechnology. 2020;**38**:1023-1033

[46] Wang XL, Yuan XZ, Huang HJ, et al. Study on the solubilization capacity of bio-oil in diesel by microemulsion technology with Span80 as surfactant. Fuel Processing Technology. 2014;**118**:141-147

[47] Chen WH, Chu YS, Liu JL, et al. Thermal degradation of carbohydrates, proteins and lipids in microalgae analyzed by evolutionary computation. Energy Conversion and Management. 2018;**160**:209-219

[48] Xu Y, Zhang L, Lv W, et al. Two-step esterification–hydrogenation of bio-oil to alcohols and esters over raney ni catalysts. Catalysts. 2021;**11**(7):818

[49] Chen WH, Wu ZY, Chang JS. Isothermal and non-isothermal torrefaction characteristics and kinetics of microalga Scenedesmus obliquus CNW-N. Bioresource Technology. 2014;**155**:245-251

[50] Li S, Zhang S, Feng Z, et al. Coke formation in the catalytic cracking of bio-oil model compounds. Environmental Progress & Sustainable Energy. 2015 [cited May 26, 2022;**34**: 240-247. DOI: 10.1002/ep.11936

[51] Saber M, Nakhshiniev B, Yoshikawa K. A review of production and upgrading of algal bio-oil. Renewable and Sustainable Energy Reviews. 2016;**58**:918-930

[52] Lian X, Xue Y, Zhao Z, et al. Progress on upgrading methods of bio-oil: A review. International Journal of Energy Research. 2017;**41**(13):1798-1816

[53] Xu J, Jiang J, Dai W, et al. Bio-oil upgrading by means of ozone oxidation and esterification to remove water and to improve fuel characteristics. Energy and Fuels. 2011;**25**:1798-1801

[54] Lødeng R, Hannevold L, Bergem H, et al. Catalytic Hydrotreatment of bio-oils for high-quality fuel production. The Role of Catalysis for the Sustainable Production of Bio-Fuels and Bio-Chemicals. Elsevier. 2013:351-396

[55] Wijayanta AT, Aziz M. Ammonia production from algae via integrated hydrothermal gasification, chemical looping, N2 production, and NH3 synthesis. Energy. 2019;**174**:331-338

[56] Shan Ahamed T, Anto S, Mathimani T, et al. Upgrading of bio-oil from thermochemical conversion of various biomass – Mechanism, challenges and opportunities. Fuel. 2021;**287**:119329

[57] Zhang S, Yang X, Zhang H, et al. Liquefaction of biomass and upgrading of bio-oil: A review. Molecules. 2019;**24**(12):2250

Chapter 7

Green Synthesis of Silver Nano-Particle from Cyanobacteria and Effect on Microalgal Growth and Production of Exopolysaccharide (EPS)

Shailendra Yadav, Shilpa Chandra, Amardeep Yadav, Avinash Kumar and Mamta Awasthi

Abstract

Cyanobacterial exopolysaccharides (EPS) are heteropolysaccharides with significant biological importance in various industries. Investigating nanoparticles is gaining interest due to their great potential in improving cyanobacterial growth and co-products accumulation. Nevertheless, green synthesis of nanoparticles offers an alternative, eco-friendly and cost-effective approach to available chemical methods of nanoparticle synthesis. Thus, this study illustrates a novel approach to green synthesizing Ag nanoparticles (AgNPs) from marine cyanobacterium *Phormidium tenue* and investigates their effect on the enhancement of biomass and exopolysaccharide accumulation in the same cyanobacterium by incorporating previously synthesized AgNPs. Firstly, the AgNPs were synthesized from *P. teneue* by adding 1 mM silver sulfate into the culture medium, and the obtained AgNPs were characterized by using UV-VIS spectroscopy, XRD, SEM, and FTIR. In order to increase the biomass yield and EPS accumulation, *P. tenue* culture was subjected to different concentrations of AgNPs. Under different concentrations of AgNPs, the biomass yield and exopolysaccharides increased compared to the control condition on the 28th and 35th day of incubation, respectively. The characterization of the obtained EPS was studied by using FTIR which showed a specific absorbance of OH, weak aliphatic C-H stretching, sulfur-containing functional groups, and carboxylic acids, revealing the characteristic feature of EPS.

Keywords: cyanobacteria, exopolysaccharides, FTIR, HPLC, XRD, SEM, UV-VIS spectroscopy, silver nanoparticle, *Phormidium tenue*

1. Introduction

1.1 Polysaccharides

Polysaccharides are biopolymers that are widely distributed in nature. Certain microorganisms have the ability to produce a large amount of polysaccharides in the presence of a surplus carbon source. Some of these polysaccharides (e.g. glycogen) serve as storage compounds while others are excreted by the cell. Different monosaccharides (hexoses and pentoses) including some complex sugars are linked glycosidically to form long chains of the polymer. These polysaccharides exhibit a wide range of chemical structures with greatly differing physical properties. A considerable expenditure of energy is incurred by the microbial cell to synthesize these biopolymers.

Microorganisms produce a diverse range of biopolymers with varied chemical properties by using both simple as well as complex substrates. These biopolymers could be either intracellular or extracellular depending upon their cellular location. Though the intracellular biopolymers are limited, nevertheless, the range of the extracellular biopolymers is vast and may be categorized into four major classes; polysaccharides, inorganic polyanhydrides (such as polyphosphates), polyesters, and polyamides to be collectively termed as extracellular polymeric substances or exopolysaccharides (EPS).

1.2 Exopolysaccharides

Exopolysaccharides are high-molecular-weight polymers that are synthesized and secreted by the microorganisms into the surrounding environment. These exopolysaccharides are mainly polysaccharidic in nature, that is, they are generally composed of monosaccharides and some non-carbohydrate substituents such as acetate, pyruvate, succinate, and phosphate. They are either covalently linked or loosely attached to the cell surface or can be released into the surrounding environment [1]. These exopolysaccharides are categorized into two groups: homopolysaccharides and heteropolysaccharides [2]. The homopolysaccharides consist of only one type of single structural unit whereas the heteropolysaccharides are composed of high-molecular-mass hydrated molecules made up of different sugar residues [3]. The composition of the EPS, however, varies with the type of microorganisms.

1.3 Cyanobacteria

In recent years, there has been a continuous search for new water-soluble polysaccharides, particularly those produced by microorganisms including cyanobacteria [4]. Cyanobacteria or blue-green algae are Gram-negative prokaryotes that perform oxygenic photosynthesis and are unicellular or filamentous. They are capable of movement by gliding when in contact with the substrate [5] and also possess the ability to survive desiccation, extremes of temperatures, high pH, and salinity [6]. They are widely distributed in diverse habitats. During their life cycle, cyanobacteria exocellularly secrete outer investments mostly constituted by heteropolysaccharides, which are frequently associated with small amounts of non-carbohydrate substituents, such as peptide, DNA, and fatty acids [7]. These exopolysaccharidic secretions

are metabolites that accumulate on the surface of microbial cells. Their presence is considered as a boundary between the microbial cell and its immediate environment serving as a barrier to successfully cope with environmental constraints against high or low temperature and salinity or against possible predators and desiccation. The production of exopolysaccharides from cyanobacteria is considered to be a good alternative for polysaccharides produced by other organisms including higher plants, bacteria, fungi etc. This is owing to the versatile nature of cyanobacteria which are able to grow in any adverse environmental conditions. Their photosynthetic mode of nutrition and simple cultural requirements further add to the convenient growth of these organisms for large-scale production. In addition, the yield of the products obtained from these organisms can be enhanced by manipulating the culture conditions [8].

1.4 Cyanobacterial EPS

There are two categories in which cyanobacterial EPS can be grouped, the first one being those which are associated with the cell surface known as cell-bound or capsular polysaccharides (CPS) and the other being the released polysaccharides (RPS) referring to those that are discharged into the surrounding environment. Depending on the thickness, consistency, and appearance, the EPS associated with the cell surface can be termed sheaths and slimes [1]. The sheath is a thin, dense layer loosely surrounding the cells or cell groups usually visible in light microscopy without staining. The slime, on the other hand, refers to the mucilaginous material dispersed around the organism but does not reflect the shape of the cells. On the contrary, the RPS is soluble aliquots of polysaccharidic material released into the medium, either from the external layer(s) or derived biosynthetically which can be easily recovered from liquid cultures.

The cyanobacterial EPS are high molecular weight complex hetero-biopolymer of 10 kDa–2 MDa. This complexity is due to the presence of branching among the monomers and frequently with other macromolecules [9]. These high molecular weight heteropolysaccharides are made up of linear or branched repeating units comprised of 2–10 monosaccharides such as hexoses, pentoses, uronic acids, and deoxy-sugars. While other important substituents include phosphate, sulfhate, acetate, pyruvate, proteins and lipids form the side chains. EPS are attached to the cell surface via hydrogen bonds, hydrophobic and electrostatic interactions.

Certain characteristic features are exhibited by the cyanobacterial EPS which are rarely found in the EPS produced by other microbial groups. For instance, the presence of uronic acid and sulfhate groups contribute to the anionic nature of the cyanobacterial EPS, conferring a negative charge and a "sticky" behavior to the overall macromolecule [1, 10]. The anionic charge plays an important role in building the affinity of these EPS towards cations, notably metal ions. Furthermore, many cyanobacterial EPS are also characterized by a significant level of hydrophobicity due to the presence of ester-linked acetyl groups, peptidic moieties and deoxysugars such as fucose and rhamnose. In the past decades, several factors controlling the production of cyanobacterial EPS have been identified. These include energy availability and the C: N ratio [11]. However, other important factors such as the effect of other nutrients as well as growth conditions such as light intensity, salinity, and temperature have not been much focused. Hence, EPS production by variation of different growth parameters becomes an important area of study.

1.5 Role of cyanobacterial EPS

Cyanobacterial EPS plays a major role in protecting cells from various stress conditions in extreme habitat by serving as boundary between the cell and the surrounding environment. EPS are considered to maintain the structure and function of the biological membrane, hence, protecting them from irreversible and lethal changes during desiccation. They possess hydrophobic/hydrophilic characteristics, owing to which they are able to trap and accumulate water; thus creating a gelatinous layer around the cell that regulates water uptake and loss and stabilizes the cell membrane during the periods of desiccation. Cyanobacterial sheath formed by EPS protects the cells from the detrimental process of biomineralization [12].

Polysaccharidic layer around the cell, in addition, prevent the cell from direct contact with toxic heavy metal present in the surrounding. Being negatively charged, these cyanobacterial EPS plays an important role in sequestration of metal cations and also create a microenvironment enriched in those metals that are essential for the growth of the cell which is otherwise present in low concentration in certain environments. The slime layer surrounding the cyanobacterial cell prevents the inactivation of nitrogenase enzyme, an enzyme responsible for nitrogen fixation which otherwise gets inhibited in presence of atmospheric oxygen. Cyanobacterial sheath also contains some UV absorbing substances such as scytonemim and mycosporine-like amino acid which protects the cell from the harmful effect of UV rays. Another important role of exopolysaccharides is that it helps in the gliding movement of cyanobacteria and also acts as an adhesive for cyanobacterial cell that lives in association or symbiosis with higher plant.

1.6 Applications of cyanobacterial EPS

Cyanobacterial exopolysaccharides possess potential applications in various fields such as food, cosmetics, environmental improvement, pharmaceutical, and water treatment industries [13, 14]. Due to the presence of both hydrophilic and hydrophobic groups in the macromolecules these exopolysaccharides act as emulsifying agent or biofloculant. Another interesting industrial application is that they have the ability to bind with the water molecules due to the presence of charged groups, finding their application in the cosmetic industry for product formulations [10]. These charged RPSs also have the capability to trap metal ions which may be used in the removal of toxic metal from polluted waters.

The most common industrial use of microbial polysaccharides is that they act as thickening agents because of their ability to modify rheological behavior of water, [15], and also to stabilize the flow properties of their aqueous solutions under drastic changes in temperature, ionic strength, and pH [1, 10]. These exopolysaccharides are water-soluble and can be used as swelling agents in the food industry due to the presence of cations such as Ca^{+2}, Fe^{+3}, Al^{+3}, Cu^{+2}, and Co^{+2}. The cyanobacterial exopolysaccharides also find their use as soil conditioners due to the N_2-fixing ability of some cyanobacterium colonies. Microbial exopolysaccharides can also be considered bioactive substances due to their possession of biological activities, such as antibacterial, anticoagulant, anti-oxidative, anticancer, and anti-inflammatory activities. This is because of the presence of sulfhate group in the molecules which interfere with the absorption and penetration of another microorganism thereby preventing or inhibiting the activity of that microorganism.

1.7 Extraction

As discussed above, whilst some EPS are tightly bound to the cell structure, others are free and directly released (RPS). Therefore, there exist some differences in their extraction methodologies. RPS can be separated using physical methods such as high-speed centrifugation and ultra-sonication whereas, firmly cells-associated EPS requires chemical methods for extraction. EPS cross-linked by divalent cations can be released from the biofilm matrix by complexing agents such as ethylenediamine tetraacetic acid (EDTA), cation-exchange resins such as Dowex or by formaldehyde treatment with or without sodium hydroxide [11].

1.8 Characterization

The monosaccharides forming the cyanobacterial biopolymers consist of many isomers and show limited absorption in UV-Vis regions making the analysis of polysaccharides very difficult in terms of detecting or identifying the macromolecule using absorbance or mass spectrometry. Total carbohydrates content can be determined by using the phenol-sulfuric method [16]. For analysis of carbohydrate composition, high-performance liquid chromatography (HPLC), however, remains the most widely used technique because of its high selectivity, sensitivity, and reliability compared to other analytical methods [7].

Though present in lower concentrations, other non-carbohydrate constituents (like protein, lipid, nucleic acid, etc.), also impart very important characteristics to the EPS due to their unique linkage to sugar moieties. Hence, the determination of these components is also of vital importance. In this regard, Fourier Transformed Infrared (FTIR) spectroscopy can be used to characterize the vibrationally active functional groups within polysaccharides.

1.9 Nanoparticles

A nanoparticle or ultrafine particle is usually defined as a particle of matter that is between 1 and 100 nm in diameter. The term is sometimes used for larger particles, up to 500 nm, or fibers and tubes that are less than 100 nm in only two directions.

Types of nanoparticles dimensions	Key features
Zero dimensions	This category includes spherical particles with diameters ranging from 1 to 100 nm Examples: nanoparticles, fuller and quantum dots
One dimensions	This category includes nanomaterials with two dimensions in the nanometer range Examples: nanotubes, nanofibers, nanowires and nanobelts.
Two dimensions	This category includes the one-dimensional nanometer range and two nanomaterials larger than 100 nm Examples: graphene, nanoscales
Three dimensions	This category, all three nanomaterial sizes are larger than 100 nm and exhibit nano-effects Example: porous nanostructure

Table 1.
Various types of nanoparticles dimensions.

Nanotechnology is the technology of inventing, synthesizing, and applying materials on the nanoscale. Nanotechnology produces materials with some specific properties and functions, which are different types from their counterparts. Nanomaterials can be classified according to their size in the range of different dimensions [17] as mentioned in **Table 1**.

Nanoparticles have several advantages over their mass particles, such as surface-to-volume ratio, which results in more significant heat treatment, better mass transfer paths, dissolution rate, and catalytic activity. In addition, nanoparticles have different functions and they are easy to function. Surface active sites have increased electrical, optical properties and improved absorption capacity.

2. Applications of nanoparticles

As the most prevalent morphology of nanomaterials used in consumer products, nanoparticles have an enormous range of potential and actual applications in agriculture, cosmetics, environment, medicine, renewable energies, and petroleum.

There are two pathways for nanoparticles preparation Klaine, *et al.*, [17],

- The direct synthetic route that produces particles in the nanosized range.
- Grinding or milling macroparticles to reduce the size.

Nanoparticles are classified in a broad spectrum according to their chemical composition, source, size, and morphology [18]. The classification of nanoparticles in the type of material, source, size, composition, and morphology.

Although nanoparticles have always existed in the environment, for example, in volcanic ash and forest fires, they are considered discoveries in the 20th century. Nanotechnology has become popular, and a variety of nanomaterials have been developed and used in various research areas. In addition, the use of nanotechnology in many industrial applications has significantly advanced technical activities. Nanoparticles are widely used in commercial products, such as plastics, cosmetics, ultra-high-resolution displays, medical applications, pharmaceuticals, the environment, etc. The application of nanotechnology and nanoparticle technology has a significant impact on the economic viability of microalgae-based products (such as oils, lipids, bioactive compounds, EPS, and biofuels).

3. Materials and method

3.1 Green synthesis of silver nanoparticles

3.1.1 Preparation of algal biomass

The cyanobacterial biomass of *P. tenue* was harvested during their exponential phase. After that, the wet algal biomass of *P. teneue* was washed thoroughly with distilled water and ultrasonicated for the green synthesis of the nanoparticle. One gram of ultrasonicated wet biomass was resuspended in 100 mL of 1 mM silver sulfate aqueous solution at pH 7 and incubated in this mixture at 25°C for 24 hours [8, 19] (**Figure 1**).

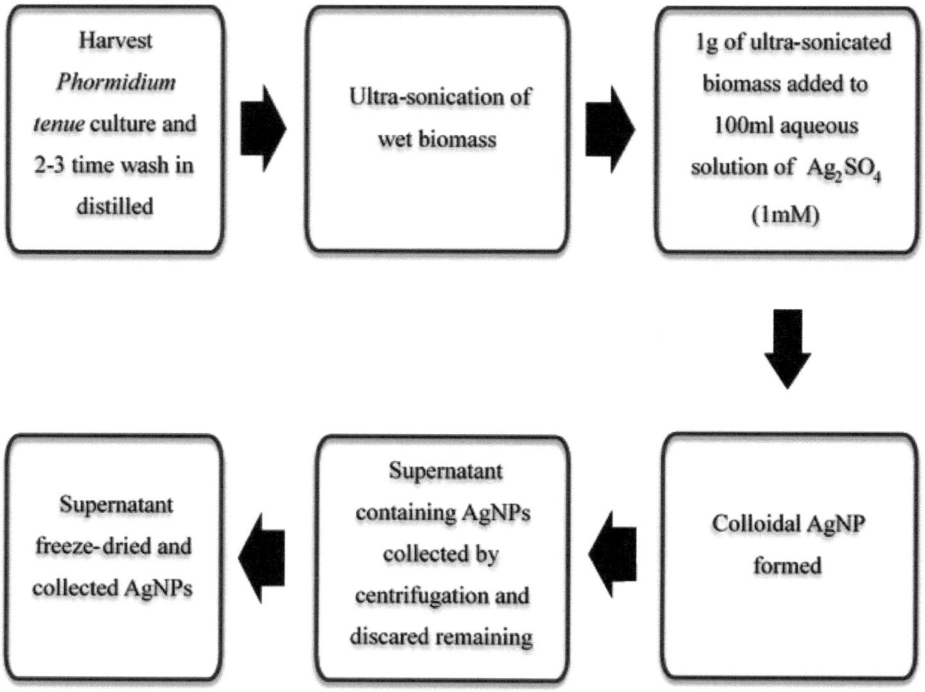

Figure 1.
Sequential step for green synthesis of silver nanoparticle from P. tenue [8, 19].

3.2 Characterization of silver nanoparticles

3.2.1 UV-Vis spectrometry

The UV-visual spectra of the synthesized nanoparticles were recorded by a UV-Vis spectrophotometer. The developed color was examined at 220–600 nm in a UV-Vis spectrophotometer (Lambda 25 UV/Vis, Perkin Elmer, Shelton, CT, USA).

3.2.2 Fourier transformer infra-red spectrometry (FT-IR)

The FTIR spectra were measured using Thermonicolet Spectrometer, Nexus 870, Thermo Nicolet, Madison, USA instrument. The synthesized green silver nanoparticle was obtained from maximum biomass culture and was pressed into KBr pellets at a ratio of 1:100. The spectra were then recorded in transmittance mode over the wave range of 4000–400 cm^{-1}.

3.2.3 X-ray diffraction analysis

The XRD analysis of the sample was collected at room temperature on a Philips X'Pert Pro diffractometer, equipped with a Cu target X-ray tube with a step size of 0.020, 2θ, and time per step of 0.3 s.

The methods of Williamson and Hall were used to calculate the crystal size and strain. The simplest and most widely used method for estimating the mean crystal

size is from the full width at half peak (FWHM) of the diffraction peak using the Scherrer equation as follows:

$$d_{XRD} = K\lambda / B\cos\theta \qquad (1)$$

Where d is the crystal size, λ diffraction wavelength, B is the corrected FWHM, is the diffraction angle, and K is the near-unit constant. The main assumption is that the sample is not deformed. B can be obtained from the observed FWHM by complicating a Gaussian configuration that models the expansion of the Br pattern, like this:

$$B^2 r = B^2{}_0 - B^2{}_i \qquad (2)$$

Where B_0 is widely observed, and Bi is the instrument broadening. Williamson and Hall is a simplified integral width method to decipher the contributions of size and strain to line expansion as a function of 2θ [20].

3.2.4 Scanning electron microscopy (SEM) analysis

The surface morphology and characteristics of the synthesized nanoparticle were observed using Scanning Electron Microscopy (SEM) according to the protocol mentioned by [21]. Images were taken by the model ZEISS SEM and performed at a beam accelerating voltage of 20 kV.

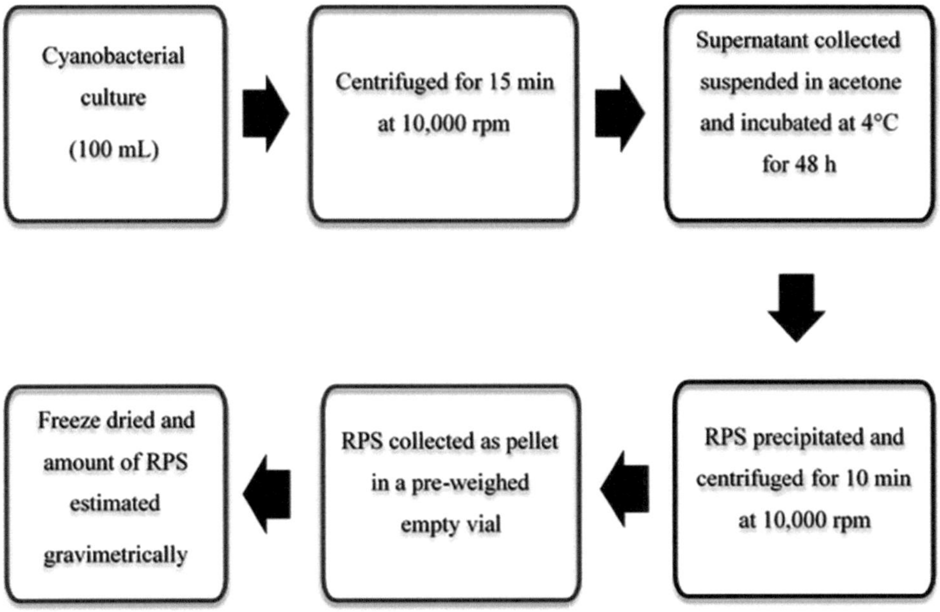

Figure 2.
Sequential step for RPS extraction from P. tenue.

3.3 Studies with the effect of silver nanoparticles

Silver nanoparticles synthesized from *P. tenue* were first filtered through a 0.22 μm membrane filter and then integrated into the control culture medium (ASN-III medium) of *Phormidium tenue* culture at different concentrations. The effects on the growth rate, biomass production, and EPS production were then monitored according to the protocol mentioned above. The total biomass yield and EPS yield were analyzed at regular intervals.

3.4 Extraction of EPS

3.4.1 Extracellular or released polysaccharides (RPS)

A known volume of culture was centrifuged at 10,000 rpm for 15 min. The supernatant so obtained was used for the extraction of released polysaccharides by addition of a measured volume of extraction solvent (acetone) followed by incubation at 4°C for 48 h. The released polysaccharide was then precipitated and collected by centrifugation at 10,000 rpm for 10 minutes. The pellet thus obtained was freeze drier (**Figure 2**).

3.4.2 Cell-bound or capsular polysaccharides (CPS)

A known amount of culture was centrifuged at 10,000 rpm for 10 min. The supernatant was discarded, and the pellet obtained was used to estimate capsular polysaccharides. It was carried out by the addition of 36.5% formaldehyde (0.06 mL) to the pellet, followed by incubation for 1 hour at 4°C [8], after which 60 mL of 1 N NaOH was introduced and further kept for incubation at 4°C for 3 hours. The treated sample was then centrifuged, and capsular polysaccharides

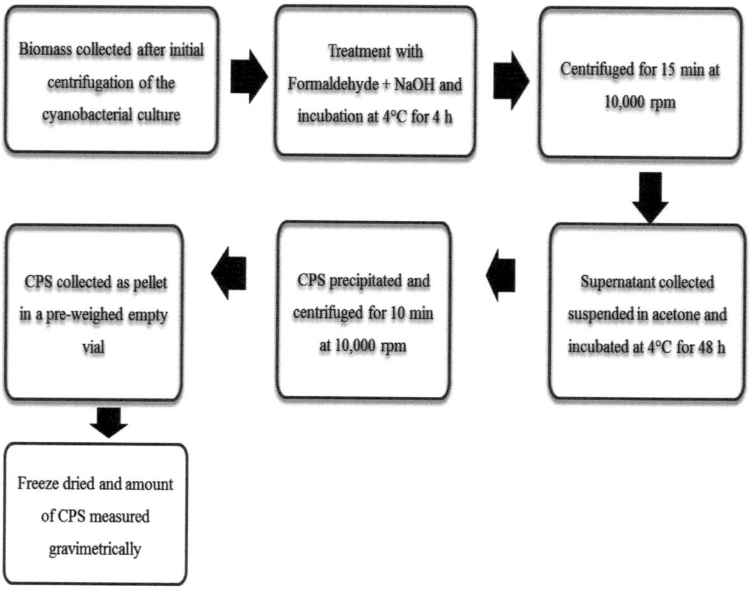

Figure 3.
Sequential step for CPS extraction from Phormidium tenue.

in that supernatant were extracted by adding acetone (10 mL) and incubated for 48 hours at 4°C. The capsular polysaccharides were precipitated by centrifugation and freeze drier (**Figure 3**).

4. Results and discussion

4.1 Green synthesis silver nanoparticle from *P. tenue*

4.1.1 UV-Vis spectrometry analysis

In this study, green synthesis of AgNP has been demonstrated from a filamentous marine *P. tenue*. It is well known that AgNPs exhibit a yellow-brown color in aqueous solution due to excitation of surface layer oscillations in AgNP. The reduction of silver ions of Silver sulfhate to AgNPs upon exposure to *P. tenue* ultrasonic biomass was followed by changing the color of the culture medium. As shown in **Figure 4A–D**, the changing color of the reaction mixture from green to yellow and then dark brown, followed by precipitation of grayish-black particles, proved the bioconversion of silver ions and the formation of AgNPs in an aqueous medium. The silver sulfate solution with washed *P. tenue* biomass turned yellow indicating the formation of silver nanoparticles.

Figure 5 shows the UV-Vis spectrum of the synthesized nanoparticle from *P. tenue*. A clear peak was observed with a maximum absorbance at 380–420 nm with

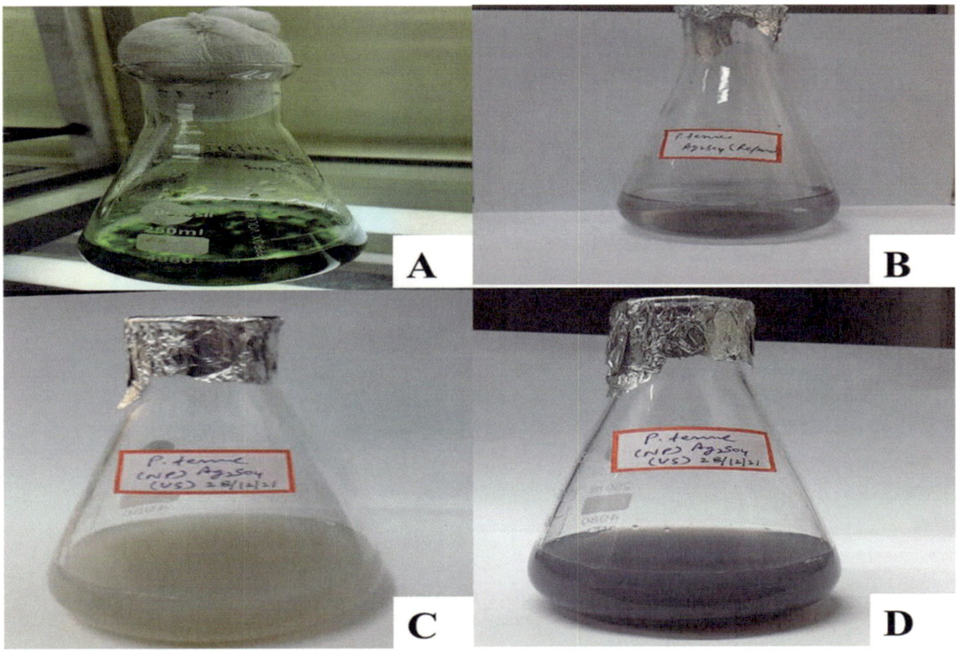

Figure 4.
(A) P. tenue culture (ASN-III medium),(B) Silver sulfate solution as the negative control, (C) Adding (1 g) ultrasonication wet biomass and Ag_2SO_4(1 Mm), the picture show the color change of silver sulfate solution by P. tenue in biomass and (D) The complete reduction of ionic silver (Ag+) and grayish black precipitation of AgNPs.

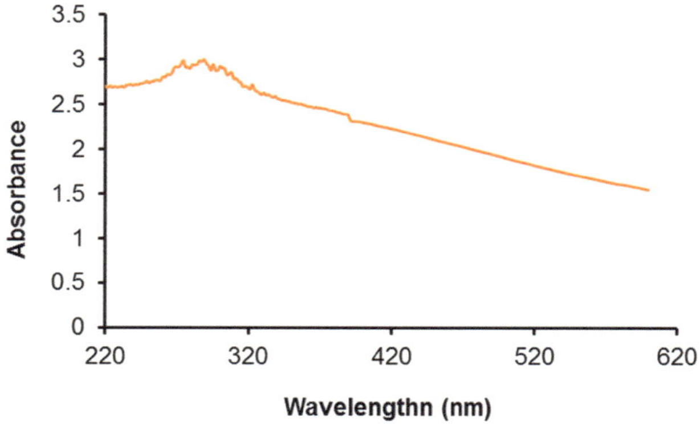

Figure 5.
UV-Vis spectrum was recorded after the reaction of 1Mm silver sulfhate solution with (1 g) P.tenue ultrasonication wet biomass at PH 7 and 25 °C and formation of AgNPs.

an absorbance of 1 mM of the silver sulfate solution. The occurrence of the peaks within shows the presence of silver nanoparticles in the solution. Agreeing Gray-black silver nanoparticle precipitation on *P. tenue* is observed in an experiment. They observed a characteristic protein coat at 270–275 nm in the ultraviolet spectrum. Ahmed et al. [22] shows that increasing the concentration of silver sulfate solution with 1 g of *Phormidium tenue* ultrasonic wet biomass causes the bioconversion of silver ions to silver nanoparticles. Furthermore, increasing the concentration of Silver sulfate solution with 1 g of *P. tenue* caused the ultrasonic wet biomass to induce the bioconversion of silver ions to silver to decrease, and the subsequent formation of SNPs in an aqueous medium. Regarding this concern, [18] observed a characteristic peak at 380–420 nm at 12 h. In principle, the wide plasma bonds with absorption at the longest wavelengths could be due to the size distribution of the nanoparticles. Silver ion reduction occurs either by an electron shuttle or by a reducing agent released by ultrasonicated *P. tenue* biomass into solution.

4.1.2 Fourier transformers infra-red spectrometry (FT-IR)

FTIR is used to identify the biomolecules in *P. tenue* responsible for the silver ions reduction and stabilization of reduced silver ions [22]. The FTIR spectrum of the AgNPs obtained from *P. tenue*, shows strong absorption peaks at 3390.90, 1634.19, 1419.41, 1111.25, 614.429, and 477.719 cm^{-1} representing different functional groups such as fragments The stretching OH of the alcohol or phenol, the N-H (amino acid), the C-O carboxylic anion, the saturated C-O group, and the stretching N-O, respectively (**Figure 6**).

The absorption peak at 3390 cm^{-1} indicates the presence of the N-H (amino acid). In agreement with this study [12] confirmed the presence of a protein coat responsible for the biosynthesis of nanoparticles. The presence of protein as a stabilizer surrounds silver nanoparticles. Protein molecule consisting of different functional groups in the amino acid chain such as amino group, carboxyl group, and sulfate group present in cyanobacterial protein promotes the formation of microscopic silver nanoparticles with narrow particle size distribution, and hydroxyl groups and sulfonic acid are beneficial for the synthesis of silver nanoparticles with slightly larger particle size in weakly reduced media.

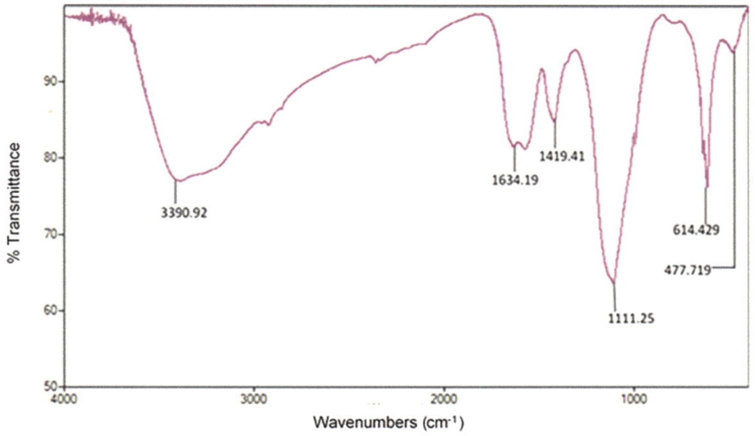

Figure 6.
FTIR analysis of Phormidium tenue show the presence of protein shell for the reduction of silver ions.

In the presence of Silver nanoparticles inside the cytoplasm, silver ions are reduced to AgNP, since Ag_2SO_4, a toxic reactant, is used in metabolism, it eventually kills the cells. When the cyanobacteria died, the silver nanoparticles produced inside the cell were released across the cell membrane into solution, as indicated by the precipitation of silver nanoparticles around the cell. The dead *P. tenue* also releases organic matter (proteins and other biochemicals), which causes silver to continue to precipitate from solution outside the cell. The protein molecules act as a reducing agent for the silver nanoparticles. Protein molecule consisting of different functional groups in the amino acid chain such as amino group, carboxyl group, and sulfate group present in cyanobacterial protein promotes the formation of silver nanoparticles. Silver ions are reduced in the presence of sulfate reductase, resulting in the formation of a stable silver hydrosol (1111.25 in cm^{-1}) and stabilized by a capping peptide [13].

4.1.3 XRD- size determination analysis

X-ray diffraction patterns have been widely used in nanoparticle research as the main characterization tool to obtain essential characteristics such as crystal structure, crystal size, and strain of nanoparticles. Randomly oriented crystals in nanocrystalline materials cause the widening of the diffraction peaks. In addition, homogeneous lattice distortion and structural defects lead to widening of peaks in diffraction patterns [23].

Figure 7 illustrates the XRD pattern of silver nanoparticles. The device apex width is obtained with standard silver powder-free from dimensional expansion, defects, and distortion. Using the Williamson and hall method and a Gaussian profile for the peak form, the average crystal sizes obtained at 60 nm and 88.18 nm for the peaks were $2\theta = 32.40$ and $2\theta = 46.40$, respectively.

4.1.4 Scanning electron microscopy (SEM) analysis

The size and structure of nanoparticles were further characterized using SEM analysis. SEM image of obtained nanoparticles clearly distinguishes the difference

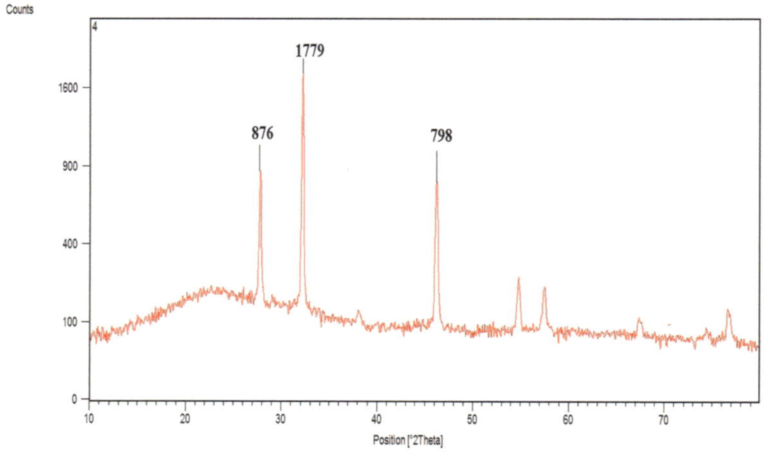

Figure 7.
The XRD pattern of silver nanoparticle.

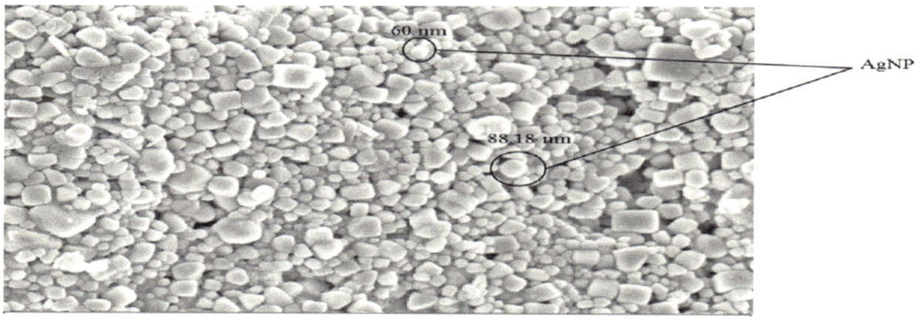

Figure 8.
SEM image of the silver nanoparticle produced by P. tenue.

between shape and size. The surface deposited silver nanoparticles are clearly seen at high magnification in the micrograph (**Figure 8**).

4.2 Estimation of exopolysaccharides (EPS) yield from supernatant of *phormidium tenue* under varying concentration of extracting solvents

EPS yield in terms of released polysaccharides (RPS) in both the cyanobacterial species using different extraction solvents viz. acetone, ethanol, and EDTA respectively taken in varying ratios with respect to the supernatant. All the three solvents gave a higher yield of EPS under supernatant: solvent ratio of 1:2 with acetone emerging out to be the best extraction solvent for cyanobacteria the test organisms.

4.3 Studies with effects of silver nanoparticles

In terms of toxicity, conducted studies have shown that Ag nanoparticles are one of the most toxic nanoparticles for microalgae due to their high reactivity, fast adsorption, and its antimicrobial properties. Thus, research effort has been directed toward finding nanoparticles that can act as nutritional supplements to increase microalgae growth and enhance the accumulation of high-value exopolysaccharides (EPS) and some other products.

5. Conclusion

The natural ability of the cyanobacteria to produce high levels of exopolysaccharide (EPS) has made them potentially attractive hosts. The present study was focused on the extraction and characterization of exopolysaccharide from cyanobacterial species namely *Phormidium tenue*. The exopolysaccharides are nothing but polysaccharide which are present on outer surface of the cell or released into the surrounding environment. The preliminary study was focused on extraction methodologies using acetone. Once the best extracting solvent was known, the studies were emphasized on the time-course analysis of the exopolysaccharide yield (released and capsular polysaccharide) from the *P. tenue* (cyanobacterial species). Green synthesis of silver nanoparticle from *P. tenue* (Cyanobacteria). The study was to characterize the silver nanoparticle through XRD, FTIR, SEM and UV-VIS spectroscopy. Later, the enhancement in microalgae growth and exopolysaccharide from *P. tenue* was observed by applying various concentrations (0.1 mg) of silver nanoparticle. Thus, the conclusion that can be drawn from the present study are:

- *P. tenue* was found to be the efficient biomass and EPS production.
- Acetone was found to be the best EPS extracting solvent in *P. tenue* (Cyanobacteria)
- Green synthesis of silver nanoparticle from *Phormidium tenue* (Cyanobacteria).
- XRD analysis of silver nanoparticle confirmed size determination by X'Pert Pro diffractometer.
- The functional groups (O-H, C=O, N-H, S=O, C-H) present in the EPS were the characteristic feature revealed by FTIR.
- Application of silver nanoparticle in enhanced biomass and EPS (both RPS and CPS) production with showing the highest biomass and EPS content than the control.

DOI: http://dx.doi.org/10.5772/intechopen.106039

Author details

Shailendra Yadav[1*], Shilpa Chandra[2], Amardeep Yadav[1], Avinash Kumar[1] and Mamta Awasthi[3]

1 Indian Institute of Technology, Kharagpur, West Bengal, India

2 Sam Higginbottom University of Agriculture Technology and Sciences, Prayagraj, Uttar Pradesh, India

3 National Institute of Technology, Hamirpur, Himachal Pradesh, India

*Address all correspondence to: shailendrayadav201291@gmail.com

References

[1] De Philippis R et al. Assessment of the metal removal capability of two capsulated cyanobacteria, Cyanospira capsulata and Nostoc PCC7936. Journal of Applied Phycology. 2003;**15**(2):155-161

[2] Sutherland IW. Novel and established applications of microbial polysaccharides. Trends in Biotechnology. 1998;**16**(1):41-46

[3] De Vuyst L, Degeest B. Heteropolysaccharides from lactic acid bacteria. FEMS Microbiology Reviews. 1999;**23**(2):153-177

[4] Singh J et al. Comparative studies of physical characteristics of raw and modified sawdust for their use as adsorbents for removal of acid dye. BioResources. 2011;**6**(3):2732-2743

[5] Bold D, Adolfo J. Atrial natriuretic factor: a hormone produced by the heart. Science. 1985;**230**(4727):767-770

[6] Flaibani A, Olsen Y, Painter TJ. Polysaccharides in desert reclamation: Compositions of exocellular proteoglycan complexes produced by filamentous blue-green and unicellular green edaphic algae. Carbohydrate Research. 1989;**190**(2):235-248

[7] Pereira L et al. A comparative analysis of phycocolloids produced by underutilized versus industrially utilized carrageenophytes (Gigartinales, Rhodophyta). Journal of Applied Phycology. **21**(5)

[8] De Philippis R et al. Exopolysaccharide-producing cyanobacteria and their possible exploitation: A review. Journal of Applied Phycology. 2001;**13**(4):293-299

[9] Rossi F, De Philippis R. Role of cyanobacterial exopolysaccharides in phototrophic biofilms and in complex microbial mats. Life. 2015;**5**(2):1218-1238

[10] Sutherland RL et al. Electrically switchable volume gratings in polymer-dispersed liquid crystals. Applied Physics Letters. 1994;**64**(9):1074-1076

[11] Liu H, Fang HHP. Characterization of electrostatic binding sites of extracellular polymers by linear programming analysis of titration data. Biotechnology and Bioengineering. 2002;**80**(7):806-811

[12] MubarakAli D et al. Plant extract mediated synthesis of silver and gold nanoparticles and its antibacterial activity against clinically isolated pathogens. Colloids and Surfaces B: Biointerfaces. 2011;**85**(2):360-365

[13] Rossi DJ et al. Cell intrinsic alterations underlie hematopoietic stem cell aging. Proceedings of the National Academy of Sciences. 2005;**102**(26):9194-9199

[14] Trabelsi L et al. Combined effect of temperature and light intensity on growth and extracellular polymeric substance production by the cyanobacterium Arthrospira platensis. Journal of Applied Phycology. 2009;**21**(4):405-412

[15] Sutherland IW. Microbial biopolymers from agricultural products: production and potential. International Biodeterioration & Biodegradation. 1996;**38**(3-4):249-261

[16] Benefield LD, Randall CW. The phenol sulfuric acid test-Effective alternative for carbohydrate analysis. Water & Sewage Works. 1976;**123**(2):55

[17] Klaine SJ et al. Nanomaterials in the environment: behavior, fate, bioavailability, and effects. Environmental Toxicology and Chemistry: An International Journal. 2008;**27**(9):1825-1851

[18] Tighe-Neira R et al. Metallic nanoparticles influence the structure and function of the photosynthetic apparatus in plants. Plant Physiology and Biochemistry. 2018;**130**:408-417

[19] Ahmed S et al. Green synthesis of silver nanoparticles using Azadirachta indica aqueous leaf extract. Journal of Radiation Research and Applied Sciences. 2016;**9**(1):1-7

[20] Akbari B, Tavandashti MP, Zandrahimi M. Particle size characterization of nanoparticles–a practicalapproach. Iranian Journal of Materials Science and Engineering. 2011;**8**(2):48-56

[21] Merin D, Devina SP, Valentine Bhimba B. Antibacterial screening of silver nanoparticles synthesized by marine micro algae. Asian Pacific Journal of Tropical Medicine. 2010;**3**(10):797-799

[22] Ahmed EA et al. Biosynthesis of silver nanoparticles by Spirulina platensis and Nostoc sp. Global Advanced Research Journal of Microbiology. 2015;**4**(4):36-49

[23] Nalwa HS. Encyclopedia of Nanoscience and Nanotechnology (v. 7. Nano Me-T). American Scientific Publishers; 2004

Chapter 8

Cyanobacteria Natural Products as Sources for Future Directions in *Antibiotic* Drug Discovery

Bahareh Nowruzi

Abstract

Cyanobacteria, an abundant source of natural products with a broad diversity of secondary metabolites, have emerged as a novel resource for the progression of synthetic analogs. Due to the rise of antibiotic resistance, there is a need for new medications and cyanobacteria-derived compounds have shown promising important alternatives for new therapeutics. These secondary metabolites are produced through nonribosomal peptide synthetase (NRPS), polyketide synthase (PKS), and mainly through mixed NRPS-PKS enzymatic systems. Current research is focused on the exploitation of cyanobacteria for the production of bioactive metabolites. Screening of cyanobacteria for pharmaceutically active compounds has received increasing attention; however, limited knowledge is available on biosynthetic mechanisms that would enhance the drug discovery process and culture-based production of desired metabolites. Overall, there is a promising outlook that cyanobacterial secondary metabolites will become alternatives for the development of new medications in a near future with enhanced pharmacological and pharmacokinetic properties.

Keywords: cyanobacteria, natural products, antibiotic, drug discovery, antibiotic resistance, polyketide synthase (PKS), nonribosomal peptide synthetase (NRPS), bioactive metabolites, synthetic analogs, biosynthetic mechanisms

1. Introduction

Antibiotics, the so-called "miracle drugs," came into existence half a century ago; however, their current popularity swiftly leads to overuse. Over the last decade, it has become quite apparent that the efficiency of antibiotics is dropping due to the growth of pathogen resistance; a problem that increases as fewer new drugs become available in the market. Moreover, unraveling this resistance is not straightforward, since antibiotic resistance is actually produced in multiple ways. Considering the urgency of the issue, efforts to develop new antibiotics are being carried out by pharmaceutical companies. In this regard, natural products account for a thorough and important component of today's pharmaceutical compendium as a fundamental source of chemical diversity. To date, several natural products have been studied, but many others still await investigation [1]. Cyanobacteria, being one of the eldest

recognized creatures living on the earth with exclusive structural features, produce several bioactive compounds with varied biological activities. Moreover, cyanobacteria as photosynthetic microorganisms, which have been preserving the oxygen levels on the earth, structurally look like gram-negative bacteria. They include chlorophyll *a* and phycobiliproteins, as well as the photosystems II and I. The adaptation mechanisms shown by cyanobacteria allow them to survive in severe climate conditions and tolerate limiting factors, such as heat, drought, salinity, nitrogen starvation, cold, photo-oxidation, osmotic, and UV stress [2]. Additionally, cyanobacteria are able to produce biologically active natural products with known antifungal, antibacterial, anti-inflammatory, antiviral, and enzyme-inhibiting bioactivities mostly through either the nonribosomal polypeptide (NRP) or the mixed polyketide-NRP biosynthetic pathways [3]. An increasing number of cyanobacterial metabolites are found to target actin and tubulin filaments in eukaryotic cells, making them a noteworthy source of anticancer natural products. Definite bioactive compounds, for example, dolastatin-10 and curacin A, have gone through clinical trials as possible anticancer drugs [4]. Cyanobacterial bioactive products can be categorized consistently with diverse structural typologies comprising terpenes, polyketides, peptides, lipids, and alkaloids. Many structural modifications can be found in cyanobacterial compounds, especially polyketide-derived units [3]. Besides, each cyanobacterial strain produces a category of bioactive compounds, so that new drugs are being constantly discovered from these sources.

Along with all these advantageous features, cyanobacteria are also known to produce toxins, mainly neurotoxins and hepatotoxins [2, 5], which act also as activators (e.g., antillatoxin) or blockers (e.g., jamaicamide A and kalkitoxin) and in addition their possible neuroprotectant and analgesics properties, they are functional molecular to distinguish usefully channels [4, 6–8].

Patellamide and trunkamide have also clinical potential, showing moderate cytotoxicity but multi-drug resistance. Investigations about the cyanobacterial natural product and secondary metabolites have gradually adapted to the genomic revolution over the past 15 years, and the genetic characterization of these secondary metabolites has led to further investigations in the field of cyanobacterial natural product synthesis. Despite important achievements in this area, numerous pharmaceutical companies have decreased the use of natural bioactive products and drug discovery screening because of: a) difficulties associated with strain, b) troubles correlated to natural bioactive products, and c) problems with logical property rights [9–14]. Finally, the use of compound collections prepared by combinatorial chemistry methods has been also influential.

2. Improving access to natural products

It is now evident that the chemical diversity of natural products is a better option than the variety of available synthetic compounds for drug discovery [15, 16]. Therefore, the use of natural chemical diversity in this regard is becoming increasingly frequent [11, 17]. Early publications showed that only a small number of cyanobacteria taxa were accessible for screening [9]. Now, extensive cyanobacteria collections, together with better cyanobacteria culture techniques, are providing new chemicals for use in drug discovery assays [11]. Progress is being made in the chemistry of natural products, leading to advances in synthetic methods seeking the production of compound analogs with enhanced pharmacological or pharmaceutical

characteristics [18]. Another interesting feature that has made natural product "privileged" structures is their ability to be used as cores of compound (alkaloids, polyketides, terpenoids, and flavonoids) libraries produced through combinatorial techniques [19, 20].

Over the past 30 years, there has been a considerable reduction in the interest by the leading pharmaceutical companies in drug discovery from natural sources. Despite this, phycologists, associated with the manufacturing industry, are exploiting this niche so that there is now a renaissance related to new improvements in spectroscopy, analytical technologies, and high-throughput screening [21]. In addition, competing technologies, such as combinatorial chemistry, have not proved to be very successful in delivering the new drug in significant numbers [22]. With the use of alternative techniques to produce analogs and derivatives of natural products, new compounds can be patented, even if the primary structure had been previously disclosed [11].

3. New approaches to the value of natural products

Multitude reasons have been suggested in regards to why natural products are such appropriate sources for drug leads, but at least one study has endeavored to quantify a connection between the drug molecules and those typically found in natural products and combinatorial chemical libraries [22]. Combinatorial libraries are synthesized in large numbers, and structures have high randomness. A multivariate evaluation of the chemical space occupied by thousands of combinatorial drug compounds compared with that of natural products revealed a good correlation between clinically approved drug molecules with natural products. This means that the structure of drugs used nowadays can be simulated by that of natural products [15]. With the progress in analytical spectroscopy, numerous clarifications are currently accessible so that the discovery of new bioactive compounds needs only a few micrograms [22]. The improvement in fractionation methods intended for isolating and purifying natural bioactive products (counter-current chromatography [20], analytical structure determination [23], etc. has led to screening natural product mixtures with timescales suitable for those expected in high-throughput screening campaigns. Complex structures can be resolved now with much less than 1 mg of the compound using the recent NMR techniques [11]. According to Quinn (developing a drug-like natural product library, 2008), it is possible to prepare a screening a library of highly diverse plant-derived compounds by pre-selecting products from the dictionary of natural products to be drug-like in their physicochemical properties. Yet, many alternative approaches are also being tested in order to enhance the speed and efficiency of drug discovery from natural products [11]. For instance, bioinformatics has been used for predicting microbes, which are able to produce new chemicals on the basis of the gene sequences encoding polyketide synthesis; this method has led to the discovery of potential antifungal and anticancer activities in some compounds [24]. Furthermore, the Metagenomics approach, which has led to the discovery of antibiotic compounds, has been recently used to achieve a broader range of synthetic cyanobacterial capabilities. This involves the collection of the entire DNA from a field cyanobacteria sample and the cloning of this DNA in host organisms, such as E. coli. Recombinant bacteria are subsequently cultured and examined for the expression of bioactive metabolites [11]. Additionally, peptide synthetase genes and polyketide synthase genes have been explored, and manipulation of biosynthetic pathways in refractory microbes, such as

uncultivable, is a promising line of research. Along with the most innovative tools of genetic engineering, new approaches to metagenomic mining of environmental DNA are being popularized, so that the genetic potential of many bacteria can be explored [25]. Even though more than 200 genome projects are either already completed or still undergoing publication, there are still some striking questions on what is actually being sequenced, considering the fact that these studies are limited to cultivable microbes. The Metagenomics approach, being culture-independent, can help to solve this problem and can also help with data mining with potential interest for a broad scientific community [25]. Different techniques unite enzymatic and synthetic methods to achieve multifaceted natural bioactive products, and refining the activity of obviously occurring antibiotics [11]. Mutasynthetic techniques are useful for making the antibiotic daptomycin-associated compounds [26, 27], vancomycin analogs and anticancer compound cryptophicin have been formed using the cytochrome P450 enzymes [12]. The biosynthesis of cyanobacterial compounds supports the creation of numerous functional groups, chiefly in the gene clusters related to cyanobacterial compounds, for instance, jamaicamide A, barbamide, or curacin A [28]. Hence, undescribed enzymatic mechanisms will be revealed thanks to biochemical studies in cyanobacterial secondary metabolic pathways. From the experience in the production of pharmaceuticals from invertebrate-derived microbes, it is evident that several obstacles must be overcome before this approach becomes a conventional technology. Still, there are good reasons to be optimistic about the future [22].

4. Activity profiling of extracts

An alternative technique to the time-consuming and expensive methods previously used for creating extensive collections of isolated and structurally characterized natural products [29] is screening the mixtures of compounds obtained from extracts of cyanobacteria strains [11]. Yet, obtaining extracts with potential biologically active novel compounds is not always simple from primary screenings. This probability can be predicted by comparing the ratio of the binding potencies at two receptor sites for a known selective ligand and for an extract by the "differential smart screens" method [30]. Furthermore, by means of a database of the usefulness of an extensive variety of identified bioactive compounds the analysis of drugs with the unknown process is imaginable. Therefore, information about previously unidentified compounds can be gained, which is precious for the antibiotic applications stated below [31]:

1. Creation of original whole-cell assays for drug screening, such as multi-patch.

2. Target identification with cDNA and quantitative real-time PCR (qRT-PCR) for confirmation of the results.

3. Revisions on mechanisms-of-action (MOA) with antibiotic-induced expression profiling.

These techniques could lead to a novel understanding of the potential effects of untested compounds (or exposure to compounds not structurally analogous and, thus, not expected to act via the same biological target) [2].

Bioinformatics and proteomics experiments are used in studies at the mRNA (transcriptome) or protein (proteome) levels, which help with the identification

of DNA binding sites of transcription factors [32] and the adjustment of biological functions, respectively, in order to characterize the complex organism responses to environmental stimulates [2]. Microarrays have been used for the identification of regulon members and stimulons by many groups in the transcriptome measurement level [33, 34].

Two-dimensional gel electrophoresis in which proteins are separated according to their molecular weight and isoelectric point, is useful in most cases, but intricate protein samples can also be analyzed using the liquid chromatography-tandem MS (LC-MS/MS) in which protein and peptide combinations are supplied to a mass spectrometer (MS) from a HPLC system. Isotopic dilution strategies on a MS instrument (e.g., isotope-coded affinity tags or ICAT) can be used for a comparative quantification of protein expression. ICAT approaches were advantageous when first released but are limited by their inability to analyze more than two conditions without a large amount of multiplexing [2, 35]. Currently, a developed version of the iTRAQ approach can analyze eight different conditions simultaneously. Despite all these tools, the most useful method would involve a concurrent quantification of the expression of all the genes and proteins of interest from a biological sample.

5. Natural products as pharmacological instruments

Aside from their curative activity, natural bioactive products can operate as pharmacological instruments demonstrating novel physiological features [14]. Cyanobacteria are stubbornly obstinate to genetic manipulation, which is accessible only for a small number of strains [3]. The modularity in cyanobacterial PKS-NRPS gene clusters authorizes the heterologous expression of natural bioactive products and, thus, genetic manipulation for combinatorial biosynthesis of innovative hybrid chemical bioactive products [4]. The prosperous production of nonribosomal and ribosomal peptides in heterologous hosts permits the usage of other cyanobacterial natural bioactive products [3]. Cyanobacteria usually synthesize multiple variants of the identical natural bioactive product; this can be ascribed to a deficiency of the inactivity of the NRPS tailoring enzymes or NRPS biosynthetic pathways. The genetic basis for this modification of secondary metabolite gene clusters is probably controlled by gene duplications, gene deletions, recombination, sequential mutation followed by natural selection, and loss and gain of tailoring enzymes [36]. However, the evolutionary and adaptive importance of these processes is deficiently understood.

6. Which cyanobacteria phyla produce therapeutics?

Throughout the prior decade, several natural bacterial compounds have been described, all of which originated from five bacterial phyla: Bacteroidetes (34 compounds), cyanobacteria (220), actinobacteria (256), proteobacteria (78), firmicutes (35), and four bioactive compounds from taxonomically unknown sources [37]. The variety of cyanobacterial natural bioactive products gathers > 1100 secondary compounds recognized with composite chemical structures, stated from different genera [3]. These metabolites represent a broad range of bioactivities including some that may be related to their natural environment (antibacterial, antifungal, antiviral, and cytotoxic) [29], but others demonstrate a

clear pharmaceutical interest, for example, they can be used as anticancer agents, immunomodulators, or protease inhibitors [38]. Cyanobacteria exhibit different growth forms, from unicellular to filamentous or colonial forms, and depending on their environmental conditions they may be surrounded by a mucilaginous or gelatinous sheath [29]. The PKS and NRPS genes seem to be more widespread in undifferentiated filamentous and heterocystous cyanobacterial strains. Despite the current taxonomic instability within cyanobacteria, which makes assessing the actual occurrence of natural products difficult, cyanobacterial compounds are mainly obtained from the lyngbya, symploca, microcystis, nostac, and hapalosiphon (**Table 1**) [3, 37].

Cyanobacterial compounds (class)	**Cyanobacterial strain**	**Biological target**	**Potential therapeutic uses**	**References**
Apratoxin A	*Lyngbya bouillonii*	STAT3, KB, and LoVo cell lines Cytotoxic against human tumor cell lines (0.36–0.52 nM)	Oncology, Early stage adenocarcinoma (induction of G-1 phase cell cycle arrest)	[4, 29, 37, 39]
Apratoxin D	*Lyngbya sp.*	Antiproliferative	Oncology	[40]
Coibamide A	*Leptolyngbya.*	Antiproliferative	Oncology	[41]
Curacin A-D NRPS-PKS	*Lyngbya majuscula 19 L*	Colon, renal, and breast cancer cell lines. Involvement of HMG-CoA in formation of cyclopropyl ring	Oncology, Antimitotic, Inhibits microtubule assembly Anti-inflammatory, Antiproliferative, Immunosuppressant, herbicidal	[4, 29, 37, 42, 43]
Cryptophycin	*Nostoc sp.*	Tubulin polymerization antiproliferative and antimitotic agents, Cytotoxicity against human tumor cell lines and human solid Tumors	Oncology, destabilization of microtubule dynamics and the induction of hyperphosphorylation of the anti-apoptotic protein B-cell leukemia/ lymphoma 2 (BCl-2),triggering programed cell death	[37, 44, 45]
Largazole	*Symploca sp.*	Histone deacetylase	Oncology, anti-epileptics, neurological disorders, mood stabilizer	[46]
Microcystin	*Microcystis aeruginosa PCC 7806,M,aeruginosa K-139 Planktothrix agardhii CYA126*	Lymphocytes	Cytotoxic, inhibit membrane-bound leucine aminopeptidase Enzyme inhibitor, cytotoxic, tumor promoter, anticancer	[47–51]
Hassallidins	*Anabaena sp. SYKE 748A*		Antifungal activity	[52]

Sulfoglycolipid	*Scytonema sp.*	HIV-1	Inhibit reverse transcriptase and DNA polymerases	[29]
Dolastatin-10	*Symploca sp.*	Binds to tubulin on rhizoxin-binding Site	Affects microtubule assembly in P388 lymphocytic leukemia cell line (NCI)	[53, 54]
Dolastatin-15	*Lyngbya sp.*	binds to vinca alkaloid site of tubulin	Breast cancers treatment	[55]
Jamaicamides (A-C)	*L. majuscula*	H-460 human lung cell carcinoma, neuro-2A-neuroblastoma cell line	Neurotoxic, cytotoxic against H-460 human lung and neuro-2a mouse neuroblastoma cell lines	[56]
Kalkitoxin	*L. majuscula*	Block voltage sensitive Na+ channel	Neurotoxic, Neural necrosis through N-methyl-D-aspartate Receptor mechanisms	[57]
Astaxanthin	*Haematococcus pluvialis*	Colon cancer cell lines	Expression decrease of cyclin D1, increase of p53 and some cyclin kinase inhibitors (p21WAF-1/CIP-1, p27)	[58]
Polysaccharide	*Navicula directa*	HSV1, 2, influenza A virus	Inhibition of hyaluronidase	[59, 60]
Allophycocyanin	*Cryptomonads*	Enterovirus 71	Inhibition of cytopathic effect, delay in synthesis of viral RNA	[61]
Hectochlorin	*L. majuscula*	Colon, melanoma, ovarian	Actin binding compounds,	[62]
Diadinochrome A, B, Diatoxanthin, cynthiaxanthin	*Peridinium bipes*	HeLa cells	Cytotoxic effect	[63]
Pheophorbide a-, b-like compounds	*Dunaliella primolecta*	HSV1	Inhibition of cytopathic effect	[64]

Table 1.
Current status of potential cyanobacteria therapeutics.

7. Cyanobacterial drug discovery

Systems biology can help us with the acquisition consciousness of the ways living systems function using computational power [65]. So as to study some specific facts in a definite biosynthetic pathway, some information about both the proteins in

charge and the responsible gene of that event is needed. The function of linking the chemical diversity of natural bioactive products and genomes in addition to modeling and prediction by incorporating such biological information could offer considerable information for the understanding of such a complex biological system [2].

According to the Comprehensive Microbial Resource Declaration, the genome sequences of human pathogenic bacteria and non-homologous in humans, have been documented. This could be an appropriate technique for the reporting of the new drug [66], and the improvements in synthetic biology now provide a solution to cyanobacteria being stubborn to genetic manipulation, opening up cyanobacteria as a valuable source of new enzymes and novel natural bioactive products.

Today pharmaceutical industry is concentrated on prominent output screening systems, genomics tools, and bioinformatics, containing combinatorial chemistry and logical design for the recognition of new bioactive compounds [29]. Recognizing groups of secondary bioactive compounds biosynthetic gene clusters with possible therapeutic competence involvement in an initial stage, which is conducted by the chemical structure of the identified bioactive compounds in cyanobacteria strains [3]. Cyanobacterial biologically active compounds are produced through NRPS, PKS, and mixed NRPS-PKS pathways [4]. Cyanobacteria strains presentation progressive screening outcomes are then designated for proteome mining and genomic characterization in order to classify biosynthetic gene clusters responsible for proteins connected to the making of these bioactive components [2]. This is imaginable because databases of biosynthetic gene clusters and cyanobacterial chemicals have been gathered through gene libraries (http://dtp.nci.nih.gov/docs/3d_database/dis3d.html, NCBI Pubchem http://pubchem.ncbi.nlm.nih.gov/, ChemIDPlus http://chem.sis.nlm.nih.gov/chemidplus, ANTIMIC [67], and Super Natural Database http://bioinformatics.charite.de/supernatural/) [68]. As a result of the increased antibiotic resistance, available drugs are effective against only one-third of the diseases, and the identification of new biologically active compounds is thus urgently necessary [29].

8. Web-based tools and databases for drug target identification

A variety of different silico tools and databases are available for drug target determination among the identified genes in pathogens for an initial screening. DrugBank(http://insilico.charite.de/supertarget/ main.html#Home), NCBI Entrez Gene(http://www.ebi.ac.uk/msd/), TarFisDock and MATADOR (http://matador.embl.de/) could be used either by a manual searching or by BLAST search of sequenced proteins. These facilities compensate the costs of screening through very large compound collections, minimizing the pace of drug discovery by both reducing the number of compounds used in real screens and the costs of screening [2].

9. Secondary metabolites derived from Cyanobacteria strains

Natural bioactive products have been isolated from a varied diversity of strains and verified for numerous biological activities. Among these strains, cyanobacteria strains signify such a source.

Secondary metabolites derived from cyanobacteria strains were identified as a rich source of bioactive compounds [69–71]. Several bioactive compounds isolated from different cyanobacterial strains showed a varied range of chemical structures

and biological activities comprising new peptides, amides, terpenes, carbohydrates, polyketides, fatty acids, alkaloids, and other organic chemicals [41, 72–74]. These compounds are regarded as good candidates for drug discovery, with functions in the industry [75–77], agriculture [19], and in pharmacy [69, 77, 78].

The cyanobacterial bioactive compounds specify useful pharmaceuticals that are problematic to produce synthetically [79]. The variety of structures found in *Lyngbya majuscula* is just incredible. Compounds isolated from this strain are amino acids, fatty acids, depsipeptides, pyrroles, amides, alkaloids, lactones, lipopeptides, and many others [40, 72, 80, 81]. Totally, cyanobacterial bioactive compounds show an exciting range of biological activities ranging from insecticidal, immunosuppressant, antiviral, anticancer, antimicrobial, and anti-inflammatory to proteinase-inhibiting activities which are outstanding targets of biomedical research (**Table 2**) [2, 5–8, 78, 113–118].

10. Antiviral activity

The extension of fatal, virus-related diseases, such as HIV, has resulted in several considerable consequences. Since the only accredited therapy (HAART, highly active antiretroviral therapy) has shown toxic effects, severe induction to viral resistance, and disability to eliminate viral agents, thus the need for new and safe antiviral therapies is an urgent issue [119, 120]. Some potential antiviral compounds are described below:

10.1 Polysaccharides

Spirulan and Ca-spirulan derived from *Spirulina sp*. are regarded as the most notable antiviral polysaccharide compounds provided their broad-spectrum activity against HIV-1, HIV-2, H, influenza and other enveloped viruses. These compounds disable the reverse transcriptase activity of HIV-1 and prevent the attachment and fusion of virus cells with host cells. Additionally, the fusion between HIV-infected and uninfected CD4+ lymphocytes, which boosts the viral infectivity, is inhibited [29]. Their reduced anticoagulant properties make them more advantageous antiviral agents over other sulfated polysaccharides. Another interesting compound is nostoflan from *Nostoc flagelliforme*, an acidic polysaccharide showing potent virucidal activity against herpes simplex virus-1 [121, 122].

10.2 Carbohydrate-binding proteins

A couple of carbohydrate-binding proteins have shown promising activity as antiviral agents. Ichthyopeptins A and B, derived from *Microcystis ichthyoblabe*, are potential agents against influenza virus, with an IC50 value of 12.5 mg ml–1 [123]. Cyanovirin-N and scytovirin are also potent virucidal drugs that interfere with several steps of the viral fusion process. Cyanovirin-N, for example, shows both *in vitro* and *in vivo* activity against HIV and other lentiviruses in nanomolar concentrations. These 101 amino acids long, 11 kDa polypeptide derived from *Nostoc ellipsosporum* is being developed as a vaginal gel for preventing sexual transmission of HIV by Cellegy Pharmaceuticals, San Francisco, CA, provided its inhibitory effects upon HIV virus-CD4 cell membrane fusion [124]. Scytovirin, on the other hand, is a 95 amino acid long, 9.7 kDa polypeptide (that includes five intra-chains disulfide bonds) derived

Species of cyanobacteria	Bioactive compounds	Biological activity	Class	References
Lyngbya Lagerheimii *Phormidium tenue*	Sulfolipid	Anti-HIV-1 activity	Fatty Acid (sulfo)	[82]
Lyngbya majuscula 19 L	Barbamide	Antimolluscicidal	chlorinated lipopeptide	[83, 84]
L. majuscule *L. majuscula*	Antillatoxin Antillatoxin B	Neurotoxic Ichthyotoxic, activator of voltage-gated sodium channel	Cyclic lipopeptide	[83, 85, 86]
Synechocystis trididemni	Didemnin	Anticancer, antiviral, immunosuppressive	Lipopeptide	[87]
Cylindrospermum licheniforme	Cylindrocyclophane	Anticancer, cytotoxic	Alkaloid Macrocycle, chloro	[43, 88, 89]
Cylindrospermopsis raciborskii	Cylindrospermopsin	Cytotoxic	Alkaloid	[90]
Prochloron didemni	Patellamide A, B, C and D	Cytotoxic, biological activity against multi-drug resistant UO-31 renal cell lines	Cyclic lipopeptide	[28, 91–93]
L. majuscula	lyngbyabellins A and B,	Cytotoxic, Anticancer, Cytoskeleton disruption	Lipopeptides	[94]
L. majuscula	Antillatoxin Antillatoxin B	Neurotoxic Ichthyotoxic, activator of voltage-gated sodium channel, with sodium channel-activating	Lipopeptide	[4, 83, 85, 86]
Lyngbya semiplena	Semiplenamides A-G	All displayed weak to moderate toxicity in brine shrimp assay; and 38 showed weak affinity for the rat cannabinoid CB1 receptor; showed moderate inhibitor of anandamide membrane transporter	Lipopeptide	[4]
Nostoc ellipsosporum	Cyanovirin	Anti-HIV, antiviral HIV-1 (interacts with high mannose groups of envelope glycoproteins, gp120 and blocks its interaction with target cell receptors) HIV-2 HSV-6 Measles virus SIV FIV	Peptide and proteins	[29, 95]
P. tenue	Monogalactopyranosyl glycerol digalactopyranosyl glycerol	Anti-HIV, anticancer	Sulfolipids	[96]

Species of cyanobacteria	Bioactive compounds	Biological activity	Class	References
Calothrix sp.	Calothrixin	Antimalarial, anticancer Against HeLa epithelial carcinoma	Indoles	[97]
Symploca hydnoides *Symploca sp VP453*	Symplostatin 1 Symplostatin 3	Against Murine colon 38 and murine mammary 16/C cell lines Against Microtubule depolymerization	analog of dolastatin-10	[54, 98]
L. majuscula	Lyngbyatoxins A-C	Cytotoxic, Ichthyotoxic, Tumor promoter, Protein Kinase C activator, Skin irritant		[2, 43]
L. majuscula	Hectochlorin	Against Colon, melanoma, ovarian and renal and lung cancer cell lines, promote actin polymerization	Lipopeptide	[4, 62]
Lyngbya majuscule	Hermitamides A and B	Ichthyotoxic, Brine shrimp toxicity, Cytotoxic		[43]
L. majuscula	Somocystinamide A	Cytotoxic against neuro-2a neuroblastoma cells (IC50 = 1.4 lg/mL), Pluripotent inhibitor of angiogenesis and tumor cell proliferation, Induces apoptosis in endothelial cells.	Lipopeptide	[4, 99]
Oscillatoria Nigroviridis	Oscillatoxin	Anticancer, Toxic general	Aromaic	[43]
Microcystis aeruginosa	Microviridin	Antibiotic, anticancer	tricyclic depsipeptides	[100, 101]
Lyngbya sp.	Kempopeptins A, B	Against colon cancer	Cyclodepsipeptides	[102]
L. majuscula	Hermitamides AeB	Against lung cancer, Potent blockers of the hNav1.2 voltage-gated sodium channel., Ichthyotoxic	Lipopeptide	[43, 103]
L. majuscula	Lyngbyatoxin A-C	Ichthyotoxic, Cytotoxic, Tumor Promoter, Protein Kinase C activator, Skin irritant		[43]
Caulerpa taxifolia Green algae	Caulerpenyne	Cytotoxicity, anticancer, antitumor, and antiproliferative activities		[104]
L. majuscula	Carmabin A-B	Anticancer, Antiproliferative	*N*-Methylated Peptide	[43]
Cyanobacteria Nostoc linckia and *Nostoc spongiaeforme var. tenue*	Boromycin	Cytotoxicity against human epidermoid carcinoma (LoVo), human colorectal adenocarcinoma activity, potent cytotoxicity against drug-resistant murine and human solid tumors	Polyether-macrolide antibiotic	[97, 105]

Species of cyanobacteria	Bioactive compounds	Biological activity	Class	References
L. majuscula	Majuscolamid A-D	Antifungal, Antimycotic activity		[106]
L. majuscula	Microcolin A-C	Antiproliferative, Anticancer, Cytotoxic, Immunosuppressive		[107]
L. majuscula	Malyngamide A-U	Antimicrobial, Antifeedant, Cytotoxic, Immunosuppressive	Lipopeptide	[43]
L. majuscula	Pitiamide A-B	Antifeedant	Fatty Acid Amides	[107, 108]
L. majuscula	Yanucamides A and B	Brine shrimp toxicity	Depsipeptides	[43]
Nodularia spumigena	Nodularia toxin	Enzyme inhibition	Lipopeptide	[2, 109, 110]
Aulosira fertilissima	Aulosirazole	Anticancer	Aromatic	[111]
Oscillatoria acutissima	Acutiphycin and 20,21-didehydroacutiphycin	Antineoplastic agent	Lipopeptide	[112]

Table 2.
Bioactive compounds from cyanobacteria.

from aqueous extracts of *Scytonema varium*, that is able to attach to the glycoprotein envelope of HIV (gp120, gp160, and gp41), thus making the virus inactive even in low nanomolar concentrations [125].

10.3 Sulfoglycolipids

Natural cyanobacterial sulfoglycolipids show confirmed HIV-reverse transcriptase and DNA polymerase inhibitory effects [29].

11. Antibacterial activity

If bacterial resistance strengthens, the treatment may become impossible for some diseases. Nosocomial infections such as those caused by the methicillin-resistant *Staphylococcus aureus* or the vancomycin-resistant enterococci, caused by multi-drug-resistant bacteria, create therapeutic problems of worldwide concern [126], hence the urgency of developing new antibiotics. Accordingly, new attempts to find antibacterial activity via screening of cyanobacterial extracts have started [127], although very few cyanobacteria-related antibacterial compounds have been detected to date. Noscomin57, from *Nostoc commune* [128], shows antibacterial activity against *Bacillus cereus, Staphylococcus Epidermidis,* and *Escherichia coli*. Antibacterial activity of Anabaena extracts against vancomycin-resistant S. aureus with a MIC of 32–64 mg ml-1 has been reported by [129].

12. Antiprotozoal activity

The estimations of the World Health Organization indicate that >109 people over the world suffer from tropical diseases caused by *Schistosoma, Trypanosoma*, *Leishmania, Plasmodium,* and others [130]. The unsuccessful treatment of such diseases (especially malaria) is related to the growing resistance shown by these protozoa and the slow pace of drug discovery [131, 132]. In a recent project operated by the Panamanian International Co-operative Biodiversity Group, five classes of antiprotozoal compounds were isolated from cyanobacteria. *Nostocarboline*, an alkaloid protease inhibitor isolated from *Nostoc sp.* 78-12 A, displayed activity against *T. cruzi, Leishmania donovani, Trypanosoma brucei,* and *Plasmodium falciparum* [133]. Moreover, aerucyclamide C68 isolated from *Microcystis aeruginosa PCC 7806* has been also detected to be active against *T. brucei*.

13. Protease inhibition activity

More than 120 cyanobacterial alkaloids with various biological activities (including protease inhibition) were introduced between 2001 and 2006. Some of these compounds, such as microginins (used for the treatment of high blood pressure), aeruginosins, and cyanopeptolins (a serine inhibitor used for asthma and viral infections) are described by Jaspars and Lawton [29]. Kempopeptins are other groups of protease inhibitory products, for example, kempopeptin B (with activity against trypsin, with an IC50 of 8.4 mM), kempopeptin A (a cyclodepsipeptide derived from marine *Lyngbya* with activity against elastase), and chymotrypsin with an IC50 of 0.32 mM and 2.6 mM, respectively [46].

14. Immunomodulatory activity

Besides the beneficial properties of cyanobacteria, their immunomodulatory activity exhibits diverse effects on immune systems, such as the increase of phagocytic activity in macrophages, the stimulation of antibody and cytokine production and the accumulation of natural killer cells into tissues, or the activation of T and B cells [134]. For instance, the effect of *Spirulina* in mice was investigated by Hayashi et al., who demonstrated increased phagocytic activity and antigen production. Enhanced phagocytic and natural killer cell-mediated antitumor activities, together with increased antigen production, were also shown in chicken by Qureshi and Ali [29]. Additionally, the incremental impact of cyanobacteria extracts on 13.6-fold interferon and 3-fold interleukin (IL)-1b and -4 was observed in human blood cells. Despite *Spirulina* has been proved to be safe, other cyanobacteria (e.g., *Microcystis sp.*) produce metabolites that are cytotoxic to lymphocytes and have inhibitory effects on membrane-bound leucine amino peptidase, which is related to antigen-processing and antigen presentation response [47, 135], confirmed the immune-toxicity of microcystin that presented medical competence in the lipopolysaccharide-induced lymph proliferation in mice vaccinated with sheep T-dependent antigen red blood cells.

15. Anticancer activity

The urgency of brand-new anticancer medications is an important issue provided the increasing resistance against currently available drugs (such as taxanes) and the outbreak of new types of cancer subjected to chemotherapeutic treatment failure [29]. A considerable number of highly active cyanobacterial compounds target tubulin or actin filaments in eukaryotic cells and have exhibited potent antimitotic properties, which makes them a noteworthy source of potential anticancer agents [4]. Several of them have gone through Phase I and II clinical trials such as the third generation dolastatin15 and TZT-1027 (soblidotin), a synthetic derivative of dolastatin-10 [136, 137]. They generally act by blocking cell division at the M-phase by targeting tubulin with efficacy equivalent to clinical drugs, such as vinblastine, vincristine, or paclitaxel. Some of these highly cytotoxic compounds are described below in **Figure 1** [138].

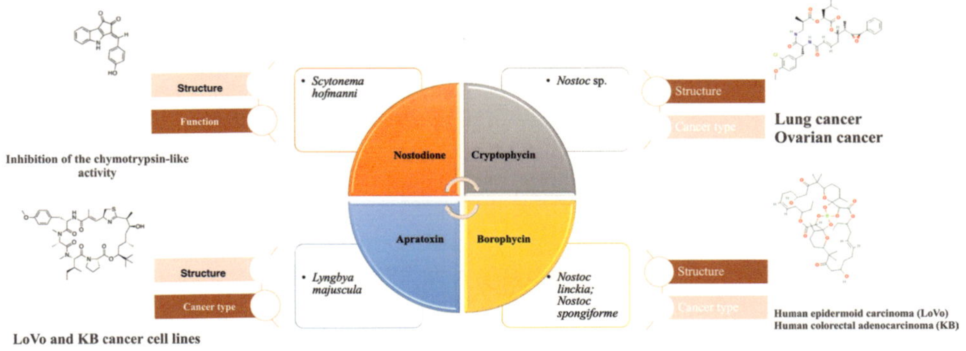

Figure 1.
Potential of cyanobacterial extract as anticancer activity.

15.1 Coibamide A – An anticancer agent with a novel action mechanism

Coibamide A, extracted from a Leptolyngbya strain, shows a novedous action mechanism targeting tubulin or actin filaments. Notable cytotoxical properties against breast, central nervous system, colon, and ovary cancers have been observed [41].

15.2 Cryptophycins

Cryptophycins are examples of cyanobacteria-derived tubulin-binding compounds with potent anticancer activity. Cryptophycin A was first isolated by Schwartz and coworkers in 1990 from Nostoc sp. strains ATCC 53789 and GSV224 [22]. Microtubule dynamics suppression and blocking of G2/M phases are features making this molecule a potent anti-carcinoma metabolite [29]. Cryptophycin-52 (LY 355073), a chemical analog of cryptophycin-1, was developed to improve its hydrolytic stability but produced very slight activity in the clinical trial. The second-generation analog, cryptophycins-249 and -309, show better water solubility and stability [139]. According to a study by [140], the thioesterase derived from the cryptophycin biosynthetic pathway through the macrocyclization of a series of linear synthetic forerunners generate 16-membered cyclic depsipeptides, showed significant efficiency as anticancer agents.

15.3 Largazole- a histone deacetylase inhibitor

Largazole, an ant proliferating compound with an unusual chemical scaffold, is extracted from *Symploca sp.* [141], and shows a considerable histone deacetylase (HDAC) inhibitory activity [142], together with a great selectivity in human mammary epithelial and fibroblastic osteosarcoma cells. The FDA ratification of HDAC inhibitor suberoylanilide hydroxamic acid as a treatment for dermal T-cell lymphomas, besides its mood stability properties and anti-epileptic characteristics, confirms this compound for cancer treatment.

15.4 Apratoxins – signal transduction inhibitors

Apratoxins, a notable class of potent cytotoxic cyclic depsipeptides, was initially isolated from a chemically rich *Lyngbya boulloni* strain and, according to NCI's Developmental Therapeutics Branch, demonstrated a unique action pattern against a panel of 60 cancer cell lines [143]. Limited findings until now indicate that the induction of G1-phase cell-cycle arrest and apoptosis is how apratoxins function as anticancer agents [39]. Apratoxin A showed moderate cytotoxicity in multiple human tumor cell lines (e.g., LoVo cell lines and KB cancer cells), although this compound is acid sensitive and decomposes when exposed to the HCl present in CDCl3. Other analogs, especially apratoxin D, have been studied in order to develop a lead structure [4].

15.5 Polypeptides- Hassallidins

Polypeptides, mostly with microbial origins, have long been used for pharmaceutical applications either as antimicrobial agents or for disinfection. A group of cyclic glycosylated lipopeptide Cyanobacteria metabolites are the hassallidins A [52] and B [144], which are purified from *Hassallia*; these compounds are a type of comprehensive with action against human pathogenic fungi [1].

16. Conclusions

The results in this review emphasized the significance of the probable healing function of natural bioactive products purified from cyanobacteria strains, for instance, antibacterial, antitumor, protease inhibition activity, and antiviral effects, and highlighted the necessity to restart discovering natural biological sources. However, system biology for metabolite purification, characterization, and valuation in cyanobacterial bioactive compounds that have not arrived in the clinical trials so far.

Author details

Bahareh Nowruzi
Department of Biology, Science and Research Branch, Islamic Azad University, Tehran, Iran

*Address all correspondence to: bahare77biol@gmail.com

References

[1] Saleem M, Nazir M, Ali MS, Hussain H, Lee YS, Riaz N, et al. Antimicrobial natural products: An update on future antibiotic drug. Natural Product Reports. 2010;**27**(2):238-254

[2] Wase NV, Wright PC. Systems biology of cyanobacterial secondary metabolite production. Expert Opinion on Drug Discovery. 1 Aug 2008;**3**(8):903-929

[3] Dittmann E, Gugger M, Sivonen K, Fewer DP. Natural Product Biosynthetic Diversity and Comparative Genomics of the Cyanobacteria. Trends in Microbiology. Cell Press. 1 Oct 2015;**23**(10):642-652

[4] Tan LT. Bioactive natural products from marine cyanobacteria for drug discovery. Phytochemistry. 1 Apr 2007;**68**(7):954-979

[5] Nowruzi B, Lorenzi AS. Production of the neurotoxin homoanatoxin-a and detection of a biosynthetic gene cluster sequence (anaC) from an Iranian isolate of Anabaena. South African Journal of Botany. 2021;**139**:300-305

[6] Ahari H, Nowruzi B, Anvar AA, Porzani SJ. The toxicity testing of cyanobacterial toxins In vivo and In vitro by mouse bioassay: A review. Mini Reviews in Medicinal Chemistry. 1 May 2022;**22**(8):1131-1151

[7] Nowruzi B, Bouaïcha N, Metcalf JS, Porzani SJ, Konur O. Plant-cyanobacteria interactions: Beneficial and harmful effects of cyanobacterial bioactive compounds on soil-plant systems and subsequent risk to animal and human health. Phytochemistry. 2021 Dec;**1**(192):112959

[8] Nowruzi B, Porzani SJ. Toxic compounds produced by cyanobacteria belonging to several species of the order Nostocales: A review. Journal of Applied Toxicology. 2021;**41**(4):510-548

[9] Baker DD et al. The value of natural products to future pharmaceutical discovery. Natural Product Reports. 2007;**24**(6):1225-1244

[10] Singh SB, Barrett JF. Empirical antibacterial drug discovery—foundation in natural products. Biochemical pharmacology. 2006;**71**(7):1006-1015

[11] Harvey AL. Natural products in drug discovery. Drug Discovery Today. 1 Oct 2008;**13**(19-20):894-901

[12] Lam KS. New aspects of natural products in drug discovery. Trends in Microbiology. 1 Jun 2007;**15**(6):279-289

[13] McChesney JD. Plant natural products: Back to the future or into extinction? Phytochemistry. 1 Jul 2007;**68**(14):2015-2022

[14] Rishton GM. Natural products as a robust source of new drugs and drug leads: Past successes and present day issues. The American Journal of Cardiology. 2008;**101**(Suppl):43D-49D

[15] Feher M, Schmidt JM. Property distributions: Differences betweendrugs, natural products, and molecules from combinatorial chemistry. Journal of Chemical. 2003;**43**:218-227

[16] Grabowski K, Schneider G. Properties and architecture of drugs andnatural products revisited. Current Chemical Biology. 2007;**1**:115-127

[17] Galm U, Shen B. Natural product drug discovery: The times have never been better. Chemistry & Biology. 2007;**14**:1098-1104

[18] Sunazuka T et al. Efficient total synthesis of novel bioactive microbial metabolites. Accounts of Chemical Research. 2008;**41**:302-314

[19] Boldi AM. Libraries from natural product-like scaffolds. Current Opinion in Chemical Biology. 2004;**8**:281-286

[20] Yao N et al. Synthesis of flavonoid analogues as scaffolds for natural product-based combinatorial libraries. Journal of Combinatorial Chemistry. 2007;**9**:668-676

[21] Jarvis LM. Liquid gold mine. Chemical and Engineering News. 2007;**85**:22-28

[22] Molinski TF, Dalisay DS, Lievens SL, Saludes JP. Drug development from marine natural products. Jan 2009;**8**(1):69-85

[23] Harvey AL. Natural products as a screening resource. Current Opinion in Chemical Biology. 2007;**11**:480-484

[24] McAlpine JB et al. Microbial genomics as a guide to drug discovery andstructural elucidation: ECO-02301, a novel antifungal agent, as an example. Journal of Natural Products. 2005;**68**:493-496

[25] Newman DJ, Hill RT. New drugs from marine microbes: The tide is turning. Journal of Industrial Microbiology & Biotechnology. 2006;**33**:539-544

[26] Kennedy J. Mutasynthesis, chemobiosynthesis, and back to semi-synthesiscombining synthetic chemistry and biosynthetic engineering for diversifying natural products. Production Reports. 2008;**25**:25-34

[27] Kopp F, Marahiel MA. Where chemistry meets biology: The chemoenzymatic synthesis of nonribosomal peptides and polyketides. Current Opinion in Biotechnology. 2007;**18**:513-520

[28] Ramaswamy AV, Flatt PM, Edwards DJ, Simmons TL, Han B. The secondary metabolites and biosynthetic gene clusters of marine cyanobacteria. Applications in biotechnology. In: Proksch P, Muller WEG, editors. Frontiers in Marine Biotechnology. 2006. pp. 175-224

[29] Singh RK, Tiwari SP, Rai AK, Mohapatra TM. Cyanobacteria: An emerging source for drug discovery. The Journal of Antibiotics. Jun 2011;**64**(6):401-412

[30] Littleton J et al. Novel approaches to plant drug discovery based on high throughput pharmacological screening and genetic manipulation. Life Sciences. 2005;**78**:467-475

[31] Freiberg C, Brotz-Oesterhelt H. The impact of transcriptome and proteome analyses on antibiotic drug discovery. Current Opinion in Microbiology. 2004;**7**(5):451-459

[32] Elnitski L, Jin VX, Farnham PJ, Jones SJM. Locating mammalian transcription factor binding sites: A survey of computational and experimental techniques. Genome Research. 2006;**16**(12):1455-1464

[33] Brötz-Oesterhelt H, Elisabeth J, Labischinski BH. Bacterial proteomics and its role in antibacterial drug discovery. Mass Spectrometry Reviews. 2005;**24**(4):549-565

[34] Pullan ST, Gidley MD, Jones RA, et al. Nitric oxide in chemostat-cultured Escherichia coli is sensed by Fnr and otherglobal regulators: Unaltered methionine biosynthesis indicates lack of S nitrosation. Journal of Bacteriology. 2007;**189**(5):1845-1855

[35] Ross PL, Huang YN, Marchese JN, et al. Multiplexed protein quantitation in Saccharomyces cerevisiae using amine-reactive isobaric tagging reagents. Molecular & Cellular Proteomics. 2004;**3**(12):1154-1169

[36] Ishida K et al. Plasticity and evolution of aeruginosin biosynthesis in cyanobacteria. Applied and Environmental Microbiology. 2009;**75**:2017-2026

[37] Williams PG. Panning for chemical gold: marine bacteria as a source of new therapeutics. Trends in biotechnology. 1 Jan 2009;**27**(1):45-52

[38] Devillers J et al. Prediction of biological activity profiles of cyanobacterial secondary metabolites. SAR QSAR Environmental Research. 2007;**18**:629-643

[39] Luesch H et al. A functional genomics approach to the mode of action of apratoxin. Natural Chemical Biology. 2006;**2**:158-167

[40] Pérez Gutiérrez RM, Martínez Flores A, Vargas Solís R, Carmona Jimenez J. Two new antibacterial norabietane diterpenoids from cyanobacteria, Microcoleous lacustris, Journal of Natural Medicines. 2008;**62**(3):328-331

[41] Medina RA et al. Coibamide A, a potent antiproliferative cyclic depsipeptide from the panamanian marine cyanobacterium Leptolyngbya sp. Journal of the American Chemical Society. 2008;**130**:6324-6325

[42] Edwards DJ, Marquez BL, Nogle LM, McPhail K, Goeger DE, Roberts MA, Gerwick WH. Structure and biosynthesis of the jamaicamides, new mixed polyketide-peptide neurotoxins from the marine cyanobacterium Lyngbya majuscula. Chemistry & Biology, 2004;**11**(6):817-833

[43] Burja AM, Banaigs B, Abou-Mansour E, Burgessd JG, Wrighta PC. Marine cyanobacteriaÐa prolific source of natural products. Tetrahedron. 2001;**57**:9347-9377

[44] Lu J, Chen F, Hodson RE. Distribution, isolation, host specificity, and diversity of cyanophages infecting marine Synechococcus spp. in river estuaries. Applied and Environmental Microbiology. 2001;**67**(7):3285-3290

[45] Mozzachiodi R, Scuri R, Roberto M, Brunelli M. Caulerpenyne, a toxin from the seaweed Caulerpa taxifolia, depresses afterhyperpolarization in invertebrate neurons. Neuroscience. 2001;**107**(3):519-526

[46] Taori K et al. Structure and activity of largazole, a potent antiproliferative agent from the Floridian marine cyanobacterium Symploca sp. Journal of the American Chemical Society. 2008a;**130**:1806-1807

[47] Mundt S, Kreitlow S, Novotony A, Effmert U. Biochemical and pharmacological investigations of selected cyanobacteria. International Journal of Environmental Health Research. 2001;**203**:327-334

[48] Christiansen G, Fastner J, Erhard M, Börner T, Dittmann E. Microcystin biosynthesis in Planktothrix: genes, evolution, and manipulation. Journal of Bacteriology. 2003;**185**(2):564-572

[49] Becker JE, Moore RE, Moore BS. Cloning, sequencing, and biochemical characterization of the nostocyclopeptide biosynthetic gene cluster: molecular basis for imine macrocyclization. Gene. 2004;**325**:35-42

[50] Welker M, Brunke M, Preussel K, Lippert I, von Döhren H. Diversity and distribution of Microcystis

(Cyanobacteria) oligopeptide chemotypes from natural communities studied by single-colony mass spectrometry. Microbiology. 2004;**150**(6):1785-1796

[51] Nowruzi B, Khavari-Nejad RA, Sivonen K, Kazemi B, Najafi F, Nejadsattari T. Identification and toxigenic potential of a cyanobacterial strain (Stigomena sp.). Progress in Biological Sciences. 2013;**3**(1):79-85

[52] Neuhof T, Schmieder P, Preussel K, Dieckmann R, Pham H, Bartl F, et al. Hassallidin A, a glycosylated lipopeptide with antifungal activity from the cyanobacterium Hassallia sp. Journal of Natural Products. 27 May 2005;**68**:695-700

[53] Horgen FD, Kazmierski EB, Westenburg HE, Yoshida WY, Scheuer PJ. Malevamide D: isolation and structure determination of an isodolastatin H Analogue from the Marine Cyanobacterium Symploca h ydnoides. Journal of Natural Products. 2002;**65**(4):487-491

[54] Luesch H, Yoshida WY, Moore RE, Paul VJ. New apratoxins of marine cyanobacterial origin from Guam and Palau. Bioorganic & Medicinal Chemistry. 2002;**10**(6):1973-1978

[55] Sato M, Nimura-Matsune K, Watanabe S, Chibazakura T, Yoshikawa H. Expression analysis of multiple dnaK genes in the cyanobacterium Synechococcus elongatus PCC 7942. Journal of Bacteriology. 2007;**189**(10):3751-3758

[56] Lautru S, Challis GL. Substrate recognition by nonribosomal peptide synthetase multi-enzymes. Microbiology. 2004;**150**(6):1629-1636

[57] LePage KT, Goeger D, Yokokawa F, Asano T, Shioiri T, Gerwick WH, Murray TF. The neurotoxic lipopeptide kalkitoxin interacts with voltage-sensitive sodium channels in cerebellar granule neurons. Toxicology Letters. 2005;**158**(2):133-139

[58] Palozza P, Torelli C, Boninsegna A, Simone R, Catalano A, Mele MC, et al. Growth-inhibitory effects of the astaxanthin-rich alga Haematococcus pluvialis in human colon cancer cells. Cancer Letters. 2009;**283**(1):108-117

[59] Catarina Guedes A, Barbosa CR, Amaro HM, Pereira CI, Xavier Malcata F. Microalgal and cyanobacterial cell extracts for use as natural antibacterial additives against food pathogens. International Journal of Food Science & Technology. 2011;**46**(4):862-870

[60] Lee JB, Hayashi K, Hirata M, Kuroda E, Suzuki E, Kubo Y, Hayashi T. Antiviral sulfated polysaccharide from Navicula directa, a diatom collected from deep-sea water in Toyama Bay. Biological and Pharmaceutical Bulletin. 2006;**29**(10):2135-2139

[61] Shih SR, Tsai KN, Li YS, Chueh CC, Chan EC. Inhibition of enterovirus 71-induced apoptosis by allophycocyanin isolated from a blue-green alga Spirulina platensis. Journal of Medical Virology. 2003;**70**(1):119-125

[62] Marquez BL, Watts KS, Yokochi A, Roberts MA, Verdier-Pinard P, Jimenez JI, et al. Structure and absolute stereochemistry of hectochlorin, a potent stimulator of actin assembly. Journal of Natural Products. 2002;**65**(6):866-871

[63] Nie B, Chen Z. Progress in studies on naturally occurring polyacetylenes. Chinese Traditional and Herbal Drugs. 1994;**11**

[64] Ohta S, Ono F, Shiomi Y, Nakao T, Aozasa O, Nagate T, et al. Anti-herpes simplex virus substances produced by the

marine green alga, Dunaliella primolecta, Journal of Applied Phycology. 1998;**10**(4):349-356

[65] H., Kitano. Systems biology: A brief overview. Science. 2002;**2002**(295):1662-1664

[66] Sakharkar KR, Sakharkar MK, VTK C. A novel genomics approach for the identifi cation of drug targets in pathogens, with special reference to pseudomonas aeruginosa. In Silico Biology. 2004;**4**(3):355-360

[67] Brahmachary M, Krishnan SPT, Koh JLY, et al. ANTIMIC: A database of antimicrobial sequences. Nucleic Acids Research. 2004;**32**(Suppl 1):D586-D589

[68] Leikoski N et al. Highly diverse cyanobactins in strains of the genus Anabaena. Applied and Environmental Microbiology. 2010;**76**:701-709

[69] Gademann K, Portmann C. Secondary metabolites from cyanobacteria: complex structures and powerful bioactivities. Current Organic Chemistry. 2008;**12**(4):326-341

[70] Smith JL, Boyer GL, Zimba PV. A review of cyanobacterial odorous and bioactive metabolites: Impacts and management alternatives in aquaculture. Aquaculture. 2008;**280**(1-4):5-20

[71] Sivonen K, Börner T. Bioactive compounds produced by cyanobacteria. In The Cyanobacteria 2008 (pp. 159-197). Caister Academic Press

[72] Tripathi A, Puddick J, Prinsep MR, Lee PP, Tan LT. Hantupeptins B and C, cytotoxic cyclodepsipeptides from the marine cyanobacterium Lyngbya majuscula. Phytochemistry. 2009;**71**(2-3):307-311

[73] Tidgewell K, Clark BR, Gerwick WH. 2.06-The natural products chemistry of cyanobacteria. In: Liu HW, Mander L, editors. Comprehensive Natural Products; 2009

[74] Portmann C, Blom JF, Gademann K, Jüttner F. Aerucyclamides A and B: isolation and synthesis of toxic ribosomal heterocyclic peptides from the cyanobacterium Microcystis aeruginosa PCC 7806. Journal of Natural Products. 2008;**71**(7):1193-1196

[75] Spolaore P, Joannis-Cassan C, Duran E, Isambert A. Commercial applications of microalgae. Journal of Bioscience and Bioengineering. 2006;**101**(2):87-96

[76] Jaiswal P, Singh PK, Prasanna R. Cyanobacterial bioactive molecules—an overview of their toxic properties. Canadian Journal of Microbiology. 2008;**54**(9):701-717

[77] Rastogi RP, Sinha RP. Biotechnological and industrial significance of cyanobacterial secondary metabolites. Biotechnology advances. 2009;**27**(4):521-539

[78] Gerwick WH, Coates RC, Engene N, Gerwick L, Grindberg RV, Jones AC, Sorrels CM. Giant marine cyanobacteria produce exciting potential pharmaceuticals. Microbe-American Society for Microbiology. 2008;**3**(6):277

[79] Kaushik P, Chauhan A. Cyanobacteria: Antibacterial Activity. New India Publishing; 2009

[80] Blunt JW, Copp BR, Munro MH, Northcote PT, Prinsep MR. Marine natural products. Natural Product Reports. 2010;**27**(2):165-237

[81] Shimizu Y. Microalgal metabolites. Current Opinion in Microbiology. 2003;**6**(3):236-243

[82] Barchi Jr JJ, Moore RE, Patterson GM. Acutiphycin and

20, 21-didehydroacutiphycin, new antineoplastic agents from the cyanophyte Oscillatoria acutissima. Journal of the American Chemical Society. 1984;**106**(26):8193-8197

[83] Orjala J, Gerwick WH. Barbamide, a chlorinated metabolite with molluscicidal activity from the Caribbean cyanobacterium Lyngbya majuscula. Journal of Natural Products. 1996;**59**(4):427-430

[84] Chang Z, Sitachitta N, Rossi JV, Roberts MA, Flatt PM, Jia J, Sherman DH, Gerwick WH. Biosynthetic pathway and gene cluster analysis of Curacin A, an Antitubulin natural product from the tropical Marine Cyanobacterium Lyngbya m ajuscula. Journal of Natural Products. 2004;**67**(8):1356-1367

[85] Yokokawa F, Fujiwara H, Shioiri T. Total synthesis and revision of absolute stereochemistry of antillatoxin, an ichthyotoxic cyclic lipopeptide from marine cyanobacterium Lyngbya majuscula. Tetrahedron. 2000;**56**(12):1759-1775

[86] Orjala J, Nagle DG, Hsu V, Gerwick WH. Antillatoxin: an exceptionally ichthyotoxic cyclic lipopeptide from the tropical cyanobacterium Lyngbya majuscula. Journal of the American Chemical Society. 1995;**117**(31):8281-8282

[87] Mitchell SS, Faulkner DJ, Rubins K, Bushman FD. Dolastatin 3 and two novel cyclic peptides from a palauan collection of Lyngbya m ajuscula. Journal of Natural Products. 2000;**63**(2):279-282

[88] Saker ML, Eaglesham GK. The accumulation of cylindrospermopsin from the cyanobacterium Cylindrospermopsis raciborskii in tissues of the Redclaw crayfish Cherax quadricarinatus. Toxicon. 1999;**37**(7):1065-1077

[89] Nakamura H, Hamer HA, Sirasani G, Balskus EP. Cylindrocyclophane biosynthesis involves functionalization of an unactivated carbon center. Journal of the American Chemical Society. 2012;**134**(45):18518-18521

[90] Carmichael WW, Eschedor JT, Patterson GM, Moore RE. Toxicity and partial structure of a hepatotoxic peptide produced by the cyanobacterium Nodularia spumigena Mertens emend. L575 from New Zealand. Applied and Environmental Microbiology. 1988;**54**(9):2257-2263

[91] Ramaswamy AV, Flatt PM, Edwards DJ, Simmons TL, Han B, Gerwick WH. The secondary metabolites and biosynthetic gene clusters of marine cyanobacteria. Applications in biotechnology. Frontiers in Marine Biotechnology. Horizon Bioscience. 2006:175-224

[92] Schmidt EW, Nelson JT, Rasko DA, Sudek S, Eisen JA, Haygood MG, Ravel J. Patellamide A and C biosynthesis by a microcin-like pathway in Prochloron didemni, the cyanobacterial symbiont of Lissoclinum patella. Proceedings of the National Academy of Sciences. 2005;**102**(20):7315-7320

[93] Williams AB, Jacobs RS. A marine natural product, patellamide D, reverses multidrug resistance in a human leukemic cell line. Cancer Letters. 1993;**71**(1-3):97-102

[94] Milligan KE, Marquez BL, Williamson RT, Gerwick WH. Lyngbyabellin B, a toxic and antifungal secondary metabolite from the marine cyanobacterium Lyngbya m ajuscula. Journal of Natural Products. 2000;**63**(10):1440-1443

[95] Esser MT, Mori T, Mondor I, Sattentau QJ, Dey B, Berger EA, Boyd MR, Lifson JD. Cyanovirin-N binds to gp120 to interfere with CD4-dependent human immunodeficiency virus type 1 virion binding, fusion, and infectivity but does not affect the CD4 binding site on gp120 or soluble CD4-induced conformational changes in gp120. Journal of Virology. 1999;**73**(5):4360-4371

[96] Hayashi O, Ono S, Ishii K, Shi Y, Hirahashi T, Katoh T. Enhancement of proliferation and differentiation in bone marrow hematopoietic cells by Spirulina (Arthrospira) platensis in mice. Journal of Applied Phycology. 2006;**18**(1):47-56

[97] Moore RE. Cyanobacterial secondary metabolites. In: Microbial Growth on C1 Compounds. Dordrecht: Springer; 1996. pp. 245-252

[98] Harrigan GG, Luesch H, Yoshida WY, Moore RE, Nagle DG, Paul VJ. Symplostatin 2: a dolastatin 13 analogue from the marine cyanobacterium Symploca hydnoides. Journal of Natural Products. 1999;**62**(4):655-658

[99] Wrasidlo W, Mielgo A, Torres VA, Barbero S, Stoletov K, Suyama TL, Klemke RL, Gerwick WH, Carson DA, Stupack DG. The marine lipopeptide somocystinamide A triggers apoptosis via caspase 8. Proceedings of the National Academy of Sciences. 2008;**105**(7):2313-2318

[100] Soria-Mercado IE, Pereira A, Cao Z, Murray TF, Gerwick WH. Alotamide A, a novel neuropharmacological agent from the marine cyanobacterium Lyngbya bouillonii. Organic Letters. 2009;**11**(20):4704-4707

[101] Han B, Gross H, Goeger DE, Mooberry SL, Gerwick WH. Aurilides B and C, Cancer Cell Toxins from a Papua New Guinea Collection of the Marine Cyanobacterium Lyngbya m ajuscula. Journal of Natural Products. 2006;**69**(4):572-575

[102] Kwan JC, Eksioglu EA, Liu C, Paul VJ, Luesch H. Grassystatins A- C from marine cyanobacteria, potent cathepsin E inhibitors that reduce antigen presentation. Journal of Medicinal Chemistry. 2009;**52**(18):5732-5747

[103] Graf KM, Tabor MG, Brown ML, Paige M. Synthesis of (S)-jamaicamide C carboxylic acid. Organic Letters. 2009;**11**(23):5382-5385

[104] Li H, Liao X, Sun Y, Zhou R, Long W, Li L, et al. An economical synthesis of caulerpin and evaluation of its new anticancer activities. ChemistrySelect. 2018;**3**(44):12406-12409

[105] Banker R, Carmeli S. Tenuecyclamides A- D, Cyclic Hexapeptides from the Cyanobacterium Nostoc s pongiaeforme var. t enue. Journal of Natural Products. 1998;**61**(10):1248-1251

[106] Mynderse JS, Hunt AH, Moore RE. 57-Normajusculamide C, a minor cyclic depsipeptide isolated from Lyngbya majuscula. Journal of Natural Products. 1988;**51**(6):1299-12301

[107] Nagle DG, Paul VJ, Roberts MA. Ypaoamide, a new broadly acting feeding deterrent from the marine cyanobacterium Lyngbya majuscula. Tetrahedron Letters. 1996;**37**(35):6263-6266

[108] Ribe S, Kondru RK, Beratan DN, Wipf P. Optical rotation computation, total synthesis, and stereochemistry assignment of the marine natural product pitiamide A. Journal of the American Chemical Society. 2000;**122**(19):4608-4617

[109] Lehtimäki J. Characterisation of cyanobacterial strains originating from the Baltic Sea with emphasis on Nodularia and its toxin, nodularin. 2000

[110] Lehtimäki J, Lyra C, Suomalainen S, Sundman P, Rouhiainen L, Paulin L, Salkinoja-Salonen M, Sivonen K. Characterization of Nodularia strains, cyanobacteria from brackish waters, by genotypic and phenotypic methods. International Journal of Systematic and Evolutionary Microbiology. 2000;**50**(3):1043-1053

[111] Bernardo PH, Chai CL, Heath GA, Mahon PJ, Smith GD, Waring P, Wilkes BA. Synthesis, electrochemistry, and bioactivity of the cyanobacterial calothrixins and related quinones. Journal of Medicinal Chemistry. 2004;**47**(20):4958-4963

[112] Reshef V, Mizrachi E, Maretzki T, Silberstein C, Loya S, Hizi A, Carmeli S. New acylated sulfoglycolipids and digalactolipids and related known glycolipids from cyanobacteria with a potential to inhibit the reverse transcriptase of HIV-1. Journal of Natural Products. 1997;**60**(12):1251-1260

[113] Nowruzi B, Haghighat S, Fahimi H, Mohammadi E. Nostoc cyanobacteria species: A new and rich source of novel bioactive compounds with pharmaceutical potential. Journal of Pharmaceutical Health Services Research. 2018;**9**(1):5-12

[114] Nowruzi B, Sarvari G, Blanco S. Applications of cyanobacteria in biomedicine. In: Handbook of Algal Science, Technology and Medicine. Academic Press; 2020. pp. 441-4553

[115] Villa FA, Lieske K, Gerwick L. Selective MyD88-dependent pathway inhibition by the cyanobacterial natural product malyngamide F acetate. European Journal of Pharmacology. 2010;**629**(1-3):140-146

[116] Abed RM, Dobretsov S, Sudesh K. Applications of cyanobacteria in biotechnology. Journal of Applied Microbiology. 2009;**106**(1):1-2

[117] Gademann K, Kobylinska J. Antimalarial natural products of marine and freshwater origin. The Chemical Record. 2009;**9**(3):187-198

[118] Dunlap WC, Battershill CN, Liptrot CH, Cobb RE, Bourne DG, Jaspars M, Long PF, Newman DJ. Biomedicinals from the phytosymbionts of marine invertebrates: A molecular approach. Methods. 2007;**42**(4):358-376

[119] Luescher-Mattli. M. Algae as a possible source of new antiviral agents. Current Medicinal Chemistry: Anti-Infective Agents. 2003;**2**:219-225

[120] Jafari Porzani S, Konur O, Nowruzi B. Cyanobacterial natural products as sources for antiviral drug discovery against COVID-19. Journal of Biomolecular Structure & Dynamics. 2021;**11**:1-7

[121] Anvar AA, Nowruzi B. Bioactive properties of spirulina: A review. Microbiology Bioactive 2021;**4**:134-142

[122] Kenji LK et al. "Isolation of an antiviral polysacharide, nostoflan, from a terrestrial cyanobacteium." Nostoc flagilliforme. Journal of Natural Products. 2005;**68**:1037

[123] Zainuddin EN et al. Cyclic depsipeptides, ichthyopeptins A and B, from Microcystis ichthyoblabe. Journal of Natural Products. 2007;**70**:1084

[124] Klasse PJ, Shattock R, Moore JP. Antiretroviral drug—Based microbicides to prevent HIV-1 sexual transmission. Annual Review of Medicine. 2008;**59**:455-471

[125] Bokesch HR et al. A potent novel anti-HIV protein from the cultured cyanobacterium. Scytonema Varium Biochem. 2003;**42**:2578-2584

[126] Reinert R, Donald EL, Rosi FX, Watal C, Dowzicky. M. Antimicrobial susceptibility among organisms from the Asia/Pacific Rim, Europe and Latin and North America collected as part of TEST and the in vitro activity of tigecycline. The Journal of Antimicrobial Chemotherapy. 2007;**60**:1018-1029

[127] Biondi N et al. Cyanobacteria from benthic mats of Antarctic lakes as a new source of bioactivities. Journal of Applied Microbiology. 2008;**105**:105-115

[128] Mundt S, Kreitlow S, Jansen R. Fatty acids with antibacterial activity from the cyanobacterium Oscillatoria redekei HUB051. Journal of Applied Phycology. 2003;**15**:263-267

[129] Bhateja P, Mathur T, Pandya M, Fatma T, Rattan A. Activity of blue-green microalgae extracts against in vitro generated Staphylococcus aureus with reduced susceptibility to vancomycin. Fitoterpia. 2006;**77**:233-235

[130] Simmons LT et al. Viridamides A and B, lipodepsipeptides with antiprotozoal activity from marine cyanobacterium. Oscillatoria Nigro Viridis. Natural Products. 2008;**71**:1544-1550

[131] Lanzer MR, Rohrbach P. Subcellular pH and Ca2+ in plasmodium falciparum implications for understanding drug resistance mechanisms. Current Science. 2007;**92**:1561-1570

[132] Sheifert K, Lunke A, Croft SL, Kayser O. Antileishmanial structure-activity relationships of synthetic phospholipids: in vitro and in vivo activities of selected derivatives. Antimicrobial Agents and Chemotherapy. 2007;**51**:4525-4528

[133] Barbaraus D, Kaiser M, Brun R, Gademann K. Potent and selective antiplasmodial activity of the cyanobacterial alkaloid nostocarboline and its dimers. Bioorganic & Medicinal Chemistry Letters. 2008;**18**:4413-4415

[134] Khan Z, Bhadouria P, Bisen PS. Nutritional and therapeutic potential of Spirulina. Current Pharmaceutical Biotechnology. 2005;**6**:373-379

[135] Shen PP et al. Effects of cyanobacteria bloom extract on some parameters of immune function in mice. Toxicology Letters. 2003;**143**:27-36

[136] Mita AC, Hammond LA, Bonate PL, Weiss G, McCreery H, Syed S, et al. Phase I and pharmacokinetic study of tasidotin hydrochloride (ILX651), a third-generation dolastatin −15 analogues, administered weekly for 3 weeks every 28 days in patients with advanced solid tumors. Clinical Cancer Research. 2006;**12**:5207-5215

[137] Watanabe J, Minami M, Kobayashi M. Antitumor activity of TZT-1027 (soblidotin). Anticancer Research. 2006;**26**:1973-1981

[138] Porzani SJ, Lima ST, Metcalf JS, Nowruzi B. In vivo and In vitro toxicity testing of cyanobacterial toxins: A Mini-review. Reviews of Environmental Contamination and Toxicology. 2021;**258**:109-150

[139] Liang J et al. Cryptophycin-309249 and other cryptophycins analogs: Preclinical efficacy studies with mouse and human tumors. Investigational New Drugs. 2005;**23**:213-224

[140] Beck ZQ, Aldrich CC, Magarvey NA, Georg GI, Sherman DH. Chemoenzymatic synthesis of cryptophycin/arenastatin natural products. Biochemistry. 2005;**44**:13457-13466

[141] Taori K, Paul VJ, Leusch H. Kempopeptins A & B, protease inhibitors with different selectivity profiles from a marine cyanobacterium Lyngbya sp. J. Natural Products. 2008b;**71**:1625-1629

[142] Ying Y et al. Total synthesis and molecular target of largazole,a histone deacetylase inhibitor. Journal of the American Chemical Society. 2008;**130**:8455-8459

[143] Federici L, Woebking B, Velamakanni S, et al. New structure model for the ATP-binding cassette multidrug transporter LmrA. Biochemical Pharmacology. 2007;**74**(5):672-678

[144] Neuhof T, Schmieder P, Seibold M. Hassallidin B—Second antifungal member of the Hassallidin family. Bioorganic and Medicinal Chemistry Letters. 2006;**16**:4220-4222